EMIL G. KELLER

E. LEITZ, INC., NEW YORK

E. LEITZ
INC
NEW YORK

THE ODYSSEY OF AN ENTERPRISE IMPORTING LEITZ SCIENTIFIC INSTRUMENTS AND LEICA CAMERAS FROM GERMANY BETWEEN 1893 AND 1980

Emil G. Keller

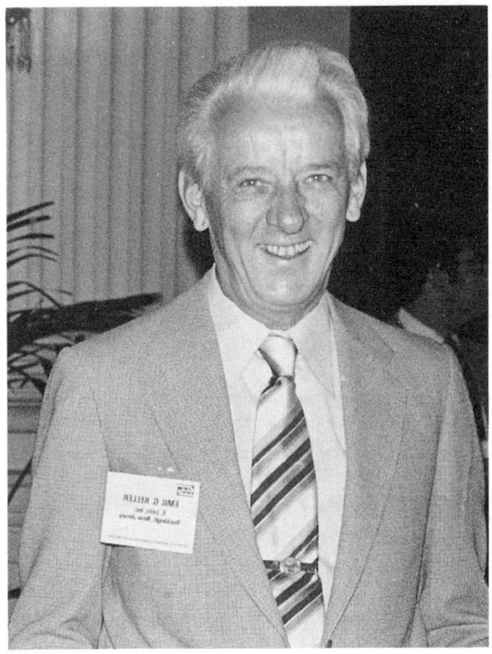

Emil G. Keller, 1976

ACKNOWLEDGEMENT

This reflection upon the importation of German Scientific Instruments and Leica Cameras to the United States could not have been accomplished without the financial assistance of the Leica Historical Society of America and Manca Inc., Westwood, New Jersey. To both my sincere thanks.

I would be amiss if I didn't also express my gratitude to Johanna for her effort, council and tolerance.

Emil G. Keller
Millbrook, New York

January 1996

For inquiries please contact:

P.O. Box 859, Millbrook, NY 12545

Tel. (914) 896-7745

LIBRARY OF CONGRESS CATALOG CARD NUMBER 96-94119
ISBN 09622612-1-1

CONTENTS

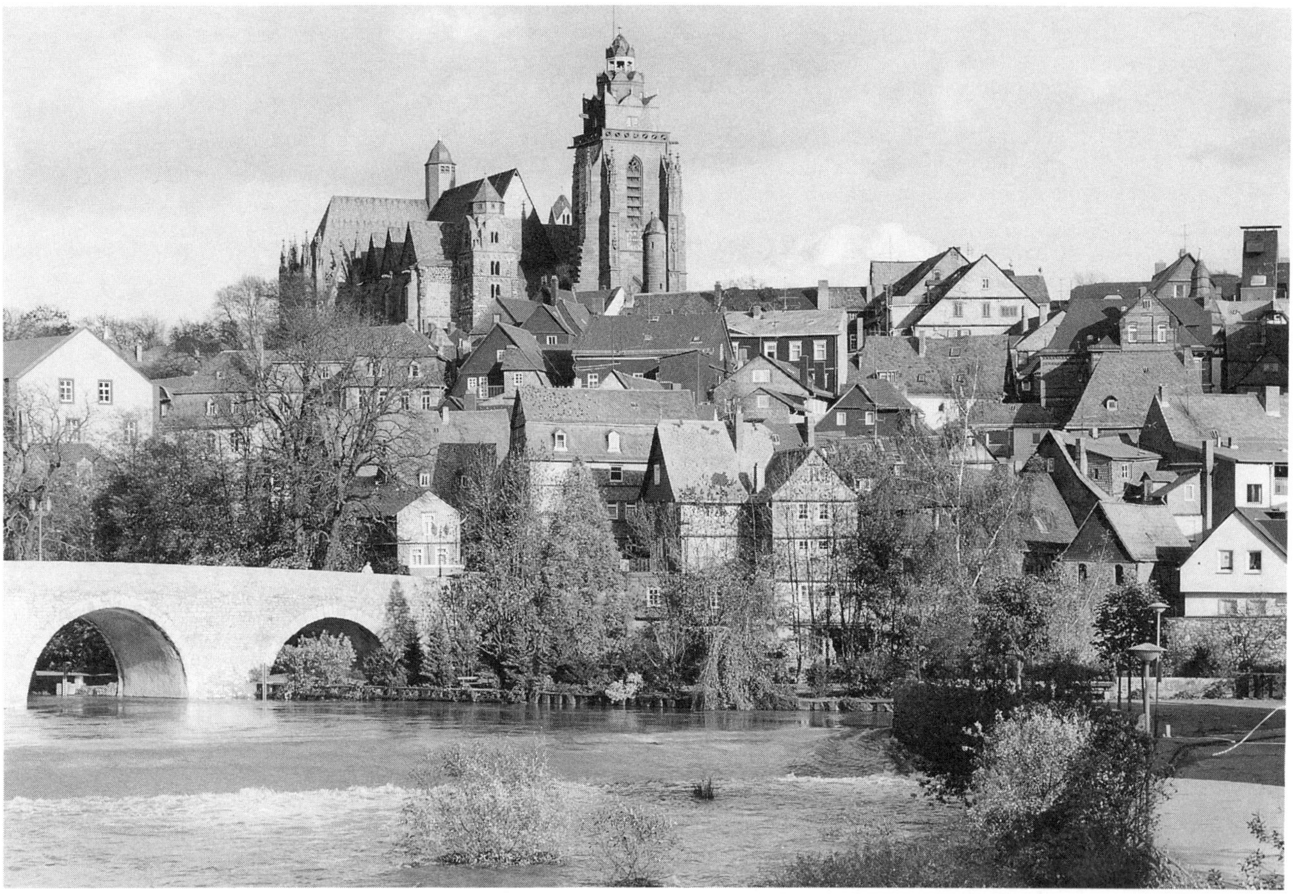

The beautiful city of Wetzler, Germany

Chapter I

DISTRIBUTORS AND IMPORTERS

Distributors and Exclusive Importers of Leitz Microscopes and Leica Cameras Manufactured by Ernst Leitz GmbH Wetzlar/Germany

To follow the fortunes of Leitz U.S.A., it becomes necessary to first comment upon the founding and development of the Leitz Works in Wetzlar, Germany.

They were started by Carl Kellner, born in 1826 in Hirzenheim, not far from Wetzlar in the State of Hesse/Rheinish Prussia. At age 21, he made his first invention, an eyepiece for field glasses and telescopes. His interests were in mathematics, physics and optics, the basics for the development of the natural sciences and medicine during the second half of the 19th century.

By 1849 he moved to Wetzlar, which at that time was a village of 5,000 persons with a history of a once celebrated court and the house where the law apprentice Goethe spent a summer.

By 1851 Kellner tried his hand at building complete microscopes which he sold for 50 to 100 Thaler (then the prevailing currency) depending upon the range of magnification desired. Kellner's assistant was Moritz Hensoldt, a friend from his school days in Hamburg. Hensoldt specialized in machinery and Kellner in optics.

Kellner passed away in 1855 at age 29 from pulmonary tuberculosis. Christian Friedrich Belthle, born in 1829, joined the firm in 1856, determined to continue Kellner's work, while also marrying Kellner's widow.

By 1867, the 1,000th microscope had been manufactured by this small firm, employing six

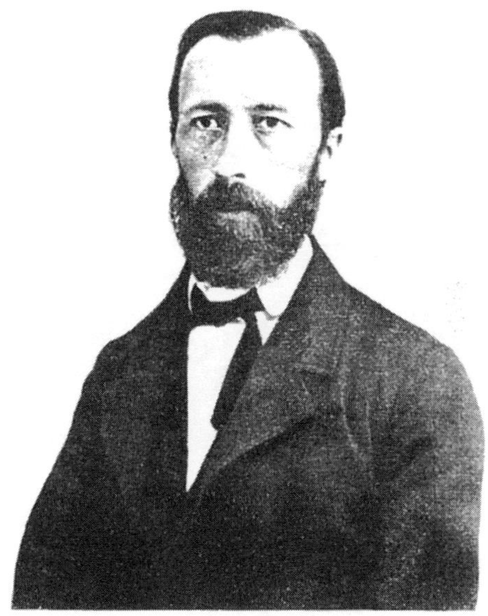

Friedrich Belthle † 1869

workers. Belthle was not a good business manager and the firm was just about holding its own when Ernst Leitz, an employee of Belthle since 1863, became his partner.

After Belthle died in 1869, Ernst Leitz took over the firm; a difficult task, while facing the Franco-Prussian War and a business less then profitable.

Ernst Leitz's father wished him to become a minister. Ernst, however, wanted to be a mechanic. He apprenticed under Christian Oechsle, a manufacturer of physical and chemical apparatus and industrial machinery in Pforzheim. At age 22, Karl Junker persuaded him to work in Switzerland

and observe the mass production methods of the Swiss watch industry. Ultimately, Leitz was to go to Wetzlar with the promise of a great opportunity in the optical and precision mechanical industry, then still in its infancy.

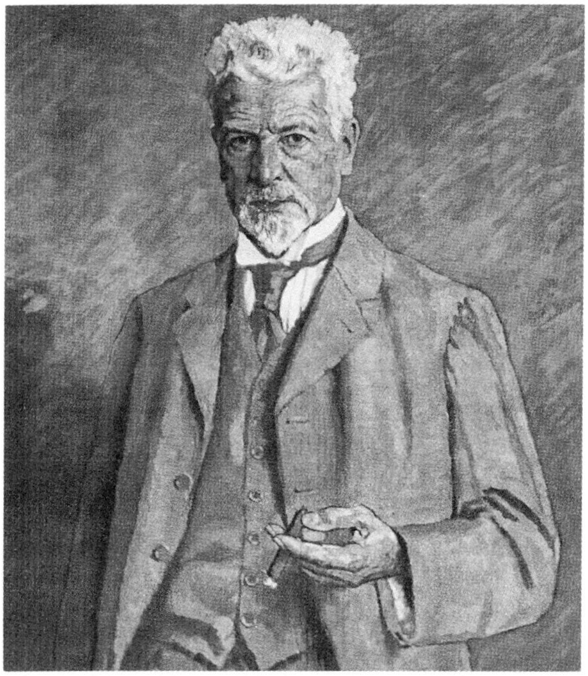

Ernst Leitz I

By 1870, the demand for microscopes had increased considerably and Leitz managed to improve the methods of manufacture, drawing from his experience in Switzerland. A most likeable personality, he made contacts with the nearby Universities of Giessen and Marburg and developed special instrumentation for the divisions of biology and medicine. He did not hesitate to even design special stands for mineralogists (1893).

The company grew and at the end of 1880 more than 2,000 microscopes per year left the factory. The number of employees reached 100 by 1887 and doubled only two years later. Competition by Carl Zeiss Jena, where Ernst Abbe had computed the first Apochromats and materially improved microscopic imaging, forced Leitz to hire the mathematician Carl Metz. Metz started a scientific department at Leitz to create new, improved optical systems.

Aside from his acumen as a businessman and a contact personality to the scientific community, Ernst Leitz also made significant technical contributions to the microscopes. He redesigned the stand itself, giving it a horseshoe foot which was later adopted by all microscope manufacturers.

The idea originated with Hartnack, a German living in Paris, whose microscopes were less expensive and more efficient. The war of 1870-71 resulted in the demise of Hartnack's Paris manufacturing facilities, increasing the demand for the German products of Leitz and Carl Zeiss Jena. It is Leitz's claim to have introduced the division of labor in microscope manufacture (previously each instrument was started and completed by one and the same mechanic). This, plus new optical systems, designed by Metz, allowed Leitz to substantially improve its market share and offer the highest quality instrumentation at competitive prices.

Meanwhile, in 1867, Ernst Leitz married a local girl, Anna Loehr, and in the same year his first son, Ludwig, was born. Ernst II followed in 1871. Anna Leitz, highly intelligent and knowledgeable soon became involved in personnel management and finances. She had been prepared for both of these disciplines by her father who owned a small local business of his own.

Some of the social amenities which Leitz introduced as early as 1885 deserve to be mentioned. First, a health insurance plan for his workers and their families. Second, a retirement program from which monies for home building were made available to employees. This was the beginning of the establishment of entire communities sponsored by the Leitz Works.

The new manufacturing facilities, established in 1875 and 1883, located away from the center of town, began to make an impact by increased output and diversity of product. In 1881, Leitz began the manufacture of microtomes. In 1907 prism binoculars became a part of the manufacturing program.

Ludwig Leitz, Ernst's first born son, followed his father's footsteps, developed apparatus for photomicrography and became the firm's first national sales manager. He visited hospitals, universities and research institutes all over the country to successfully promote Leitz microscopes.

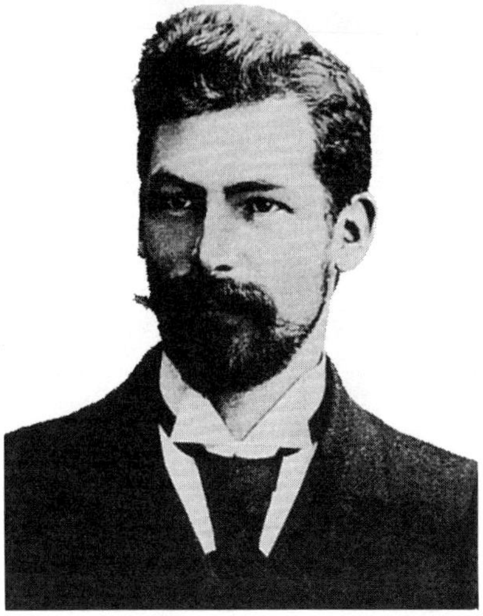

Ludwig Leitz
the first born son of Ernst Leitz I

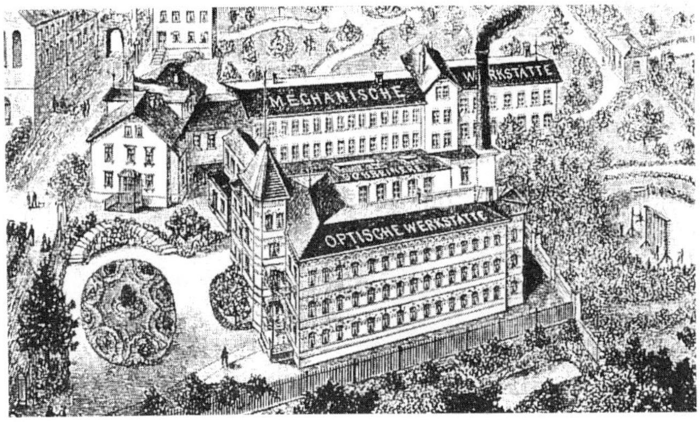

Leitz factory view 1890

Photo-micrographic Apparatus.

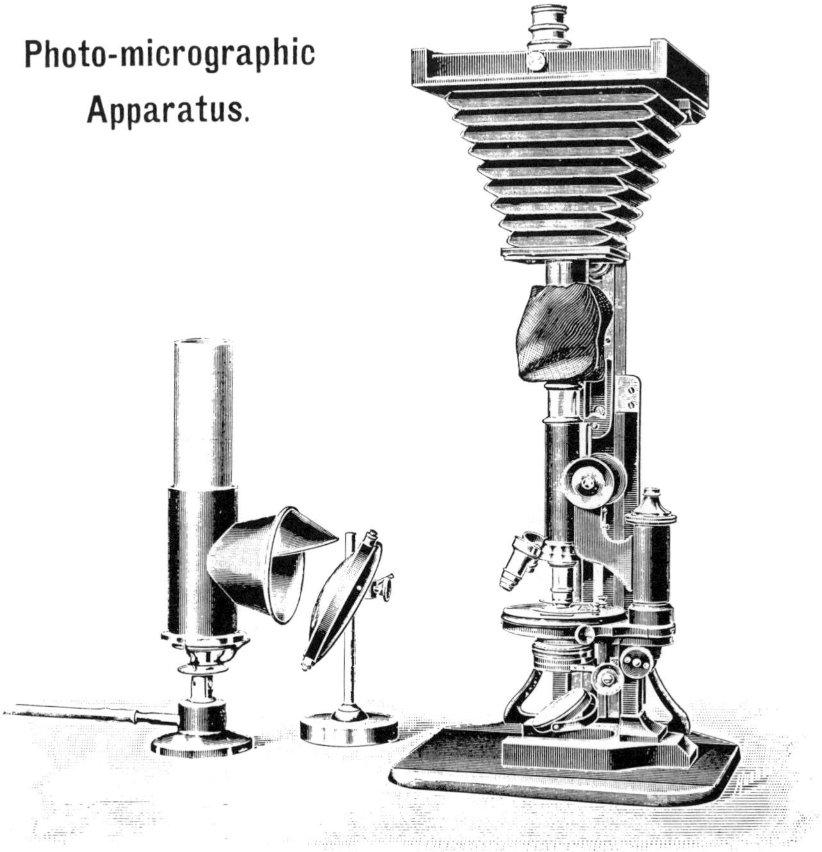

Photomicrographic apparatus (1890)

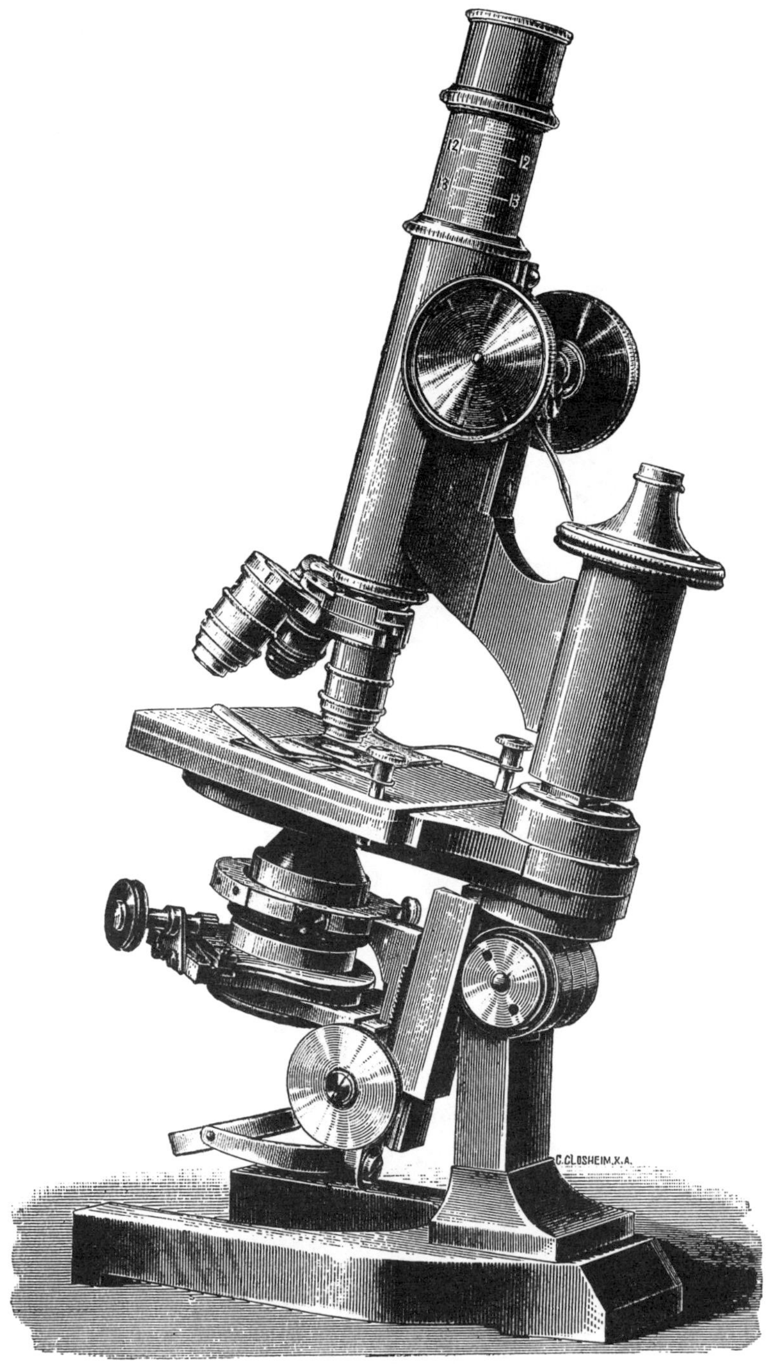

Leitz's No.1 Stand
for zoology, botany and general microscopy

Chapter 2

THE FOUNDING OF LEITZ U.S.A.

The Founding of Leitz U.S.A.

The New York branch of Ernst Leitz GmbH, Wetzlar/ Germany, was founded in 1893 by Ludwig Leitz, the eldest son of Ernst Leitz the First. Ludwig Leitz came to the United States together with William Krafft, a Wetzlar-born individual with experience in foreign trade. The first address (1893-94) of the new enterprise was listed under the name of "William Krafft, Microscopes, Importer", at 30 East 23rd Street, New York, N.Y. In 1903, the Company took the name "Ernst Leitz Microscopes", still using the same address.

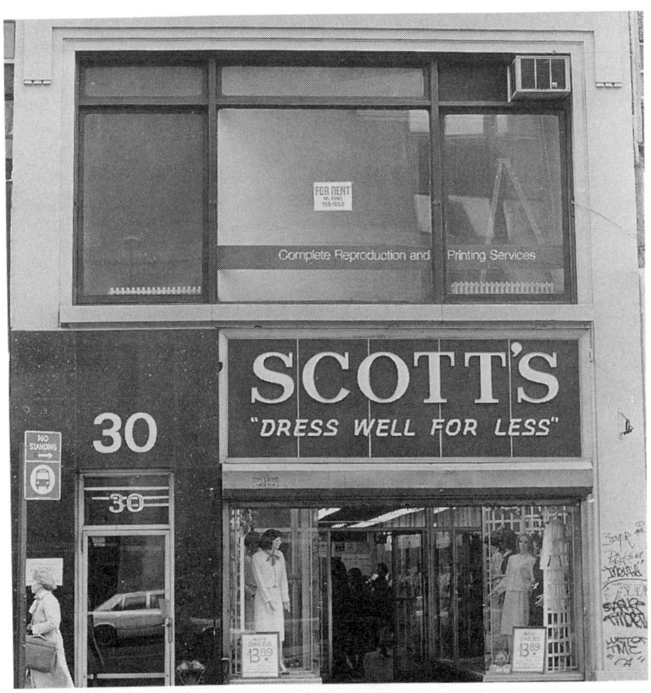

30 East 23rd Street
New York City-today

While Ludwig Leitz returned to Germany, the management of the new branch was placed into the hands of William Krafft, who operated it until 1908. Ludwig Leitz died in 1896, as the result of a riding accident. He fell off his horse in front of the factory, suffering severe head injuries from which he ailed for four years without recovering.

From 1904 to 1906, the business of Ernst Leitz Microscopes listed 411 West 59th Street as its address in New York. It remained there till 1920.

In 1906, William Krafft returned to Wetzlar. On his way back from New York he visited the major universities in England and, at the behest of Leitz Wetzlar, founded a branch office in London, England.

A few years later, Krafft became ill and returned to Germany for health reasons. Charles Rheinhold was entrusted with the Leitz U.S.A. management position by Krafft, but since Rheinhold did not have any technical training, Ernst Leitz II appointed Alfred Traeger to the position of manager of the New York agency. Traeger arrived in New York in 1910 together with Henri Dumur. Dumur studied American production methods in the optical industry with such firms as Bausch & Lomb in Rochester, New York, and Spencer Lens Company in Buffalo, New York. He did not participate in the operation of Leitz U.S.A. Dumur's position at that time was that of an observer, a friend and distant relative of the family.

Alfred Traeger began his career in the optical industry at Hansoldt & Sons, Wetzlar, manufacturers of field glasses for hunters and microscopes. Traeger was largely educated in the pursuit of commerce, short hand, typing, English and French. His position at Hansoldt was that of head of Ordering and Correspondence Departments.

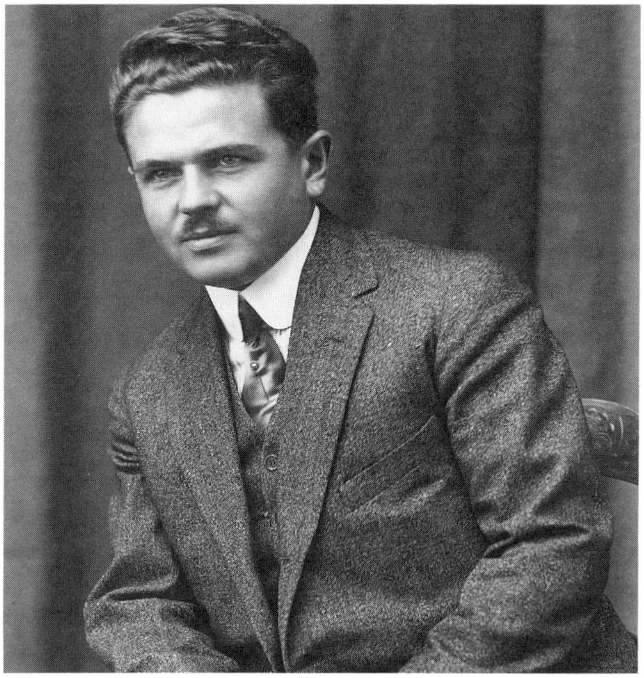

Alfred Traeger

Meanwhile, Ernst Leitz II, at age 32 in 1903, became a partner in his father's firm, where he diversified production while remaining true to his concern for the well being of his work force. The spirit of cooperation was the key to success. The Leitz Works were already recognized the world over as a key enterprise in the advancing technological society. His father, the founder, died in 1920 at age 77.

Ernst Leitz and Alfred Traeger met at the local rowing club in Wetzlar, an important social meeting place where fitness and comradery was fostered. Ernst Leitz II, more than 6 ft. tall, slender and an enthusiastic rower, was a member of that club. Traeger, too, joined the rowing fraternity and soon a friendship developed. Ultimately, in 1907, Traeger left Hansoldt & Sons and joined the Leitz Company. He underwent technical training and was appointed to the position of manager of the Leitz Frankfurt branch.

Traeger was sent to the U.S.A. because that Leitz agency had been neglected and was financially less than self supporting. The branch needed strong measures to become profitable again, and Traeger was the best man for the job. He landed in New York on February 10th, 1910.

At the end of the nineteenth century, German scientists were at the leading edge of medical discoveries, in which the microscope played an important role. Bacteriologists, pathologist and biologists used microscopes to advance their respective sciences. Equally important were new methods in the field of metallurgy as well as ore microscopy. Again, Leitz pioneered instrumentation based upon recommendations by the scientific-technical community. Such instruments found ready markets in the U.S.A.

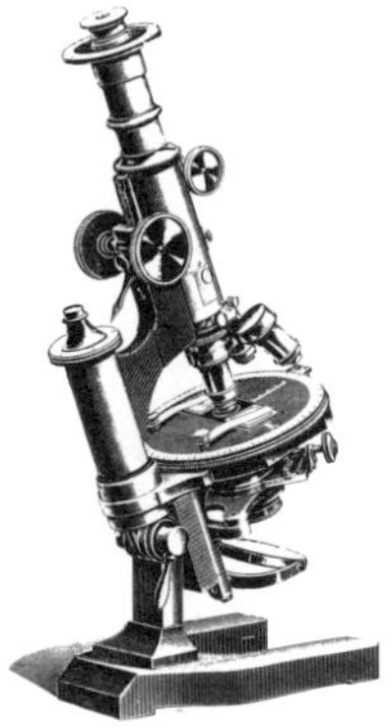

Mineralogical Stand I.

The "New World" was still looked upon by most Europeans as somewhat uncertain business

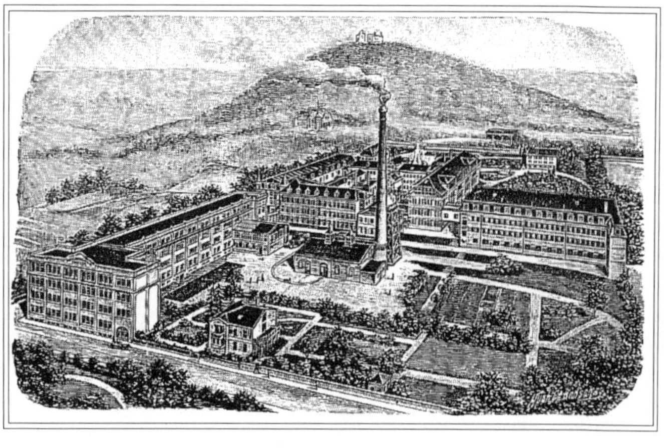

Leitz Optical and Mechanical Works (1910)

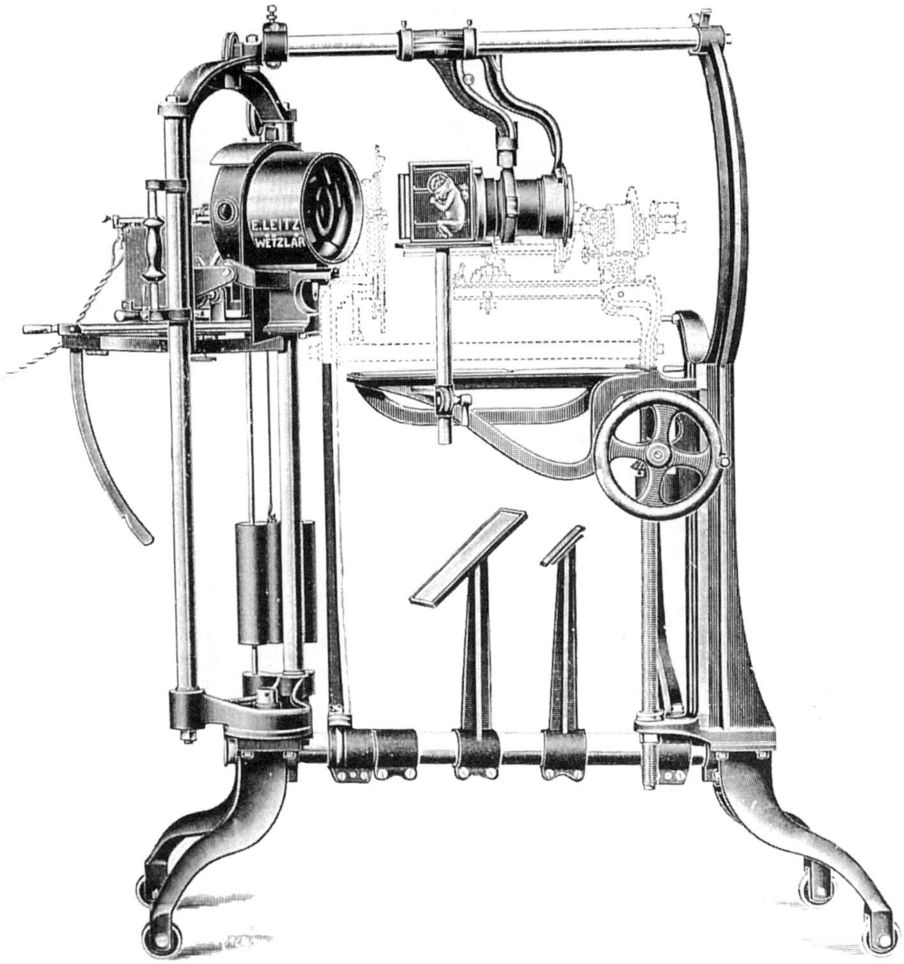

Kaiserling Universal Projection Apparatus (1910)

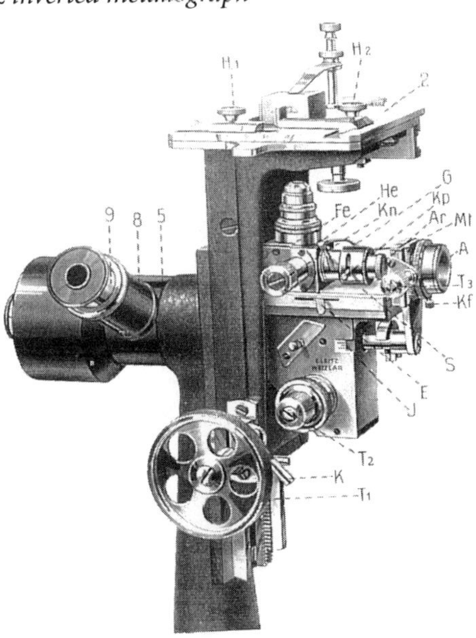

Leitz inverted metallograph

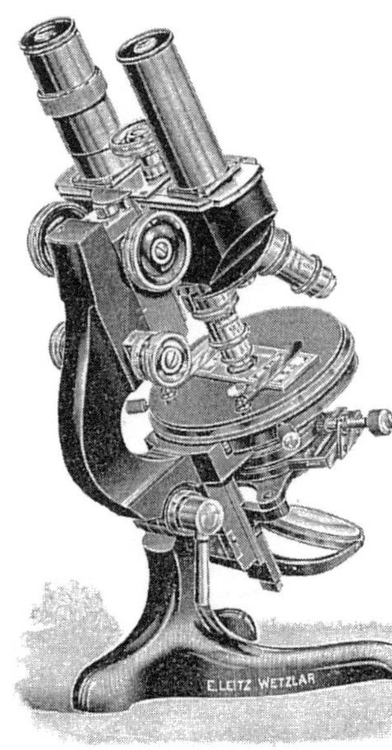

Leitz binocular microscope (1910)

territory. Wetzlar's decision to do business in the U.S.A. was based on self-confidence and assurance in the quality of their product. New York was selected because it was (and still is) the most important city in the U.S. Early catalogs (1909 - 1910) indicate that Leitz sold in the U.S. an inverted microscope according to the "Le Chatelier" principle. This instrument was used for the examination of ores and metals combined with facilities to photograph these specimen under more than 1000x magnification on 8" x 1O" plate cameras.

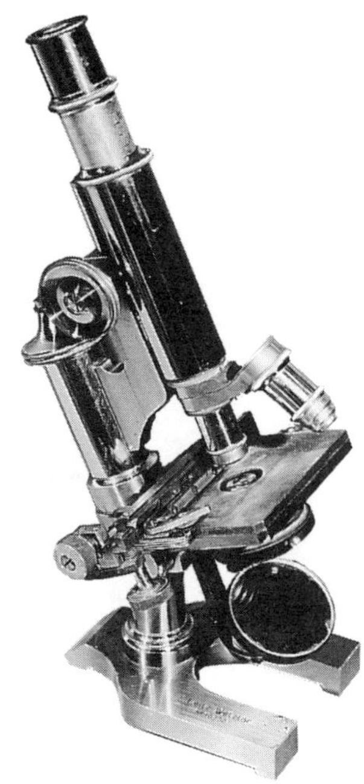

Microscope No. 40752 (1898)

This was also the time when Leitz introduced a new binocular microscope. "Ever since optical instruments were known, people have tried to make them suitable for use with both eyes", as stated in an article by Felix Jentsch, PhD, on December 17, 1913, in the Journal of the Royal Microscopical Society. The problem was to construct a binocular microscope which may be used with any desired eyepiece and in which the two fields of vision have equal light intensity and all objectives, including oil immersion, may be used. This problem was solved with the new binocular microscope by Leitz.

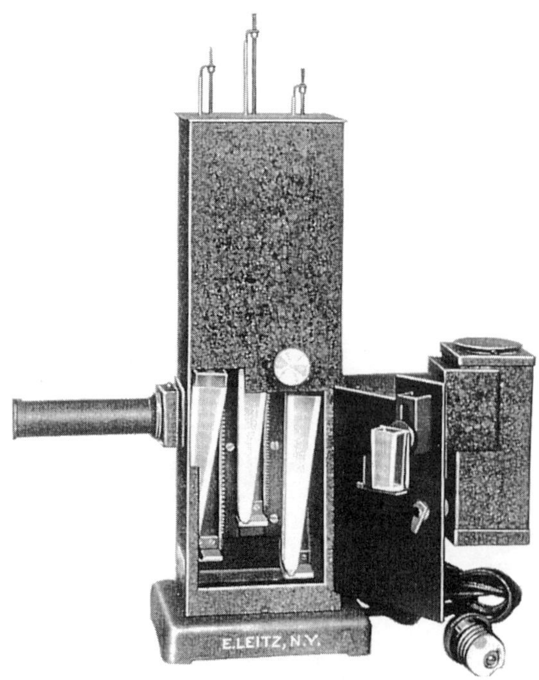

Bi-Colorimeter after Meyers

In 1914, Dr. Ernst Leitz II travelled to New York on the steamer "Vaterland" of the Hapag America Line to visit his growing sales agency. He also sought to obtain details about the American Film Industry, especially with respect to film projection apparatus. Meschau, one of Leitz's engineers, designed a projector that promised to eliminate the wear and tear on motion picture films, while at the same time steadying the image on the screen. He also brought with him another invention, made by a Wetzlar engineer named Barnack, years later to be called the "Leica" (1924). Leitz later introduced this as "the camera of the future".

Under paragraphs 573 and 654 of the U.S. Tariff Act of 1913, apparatus of foreign origin could be imported for educational institutions free of U.S. Customs Duties, subject to regulations by the Treasury Department. Leitz New York, therefore,

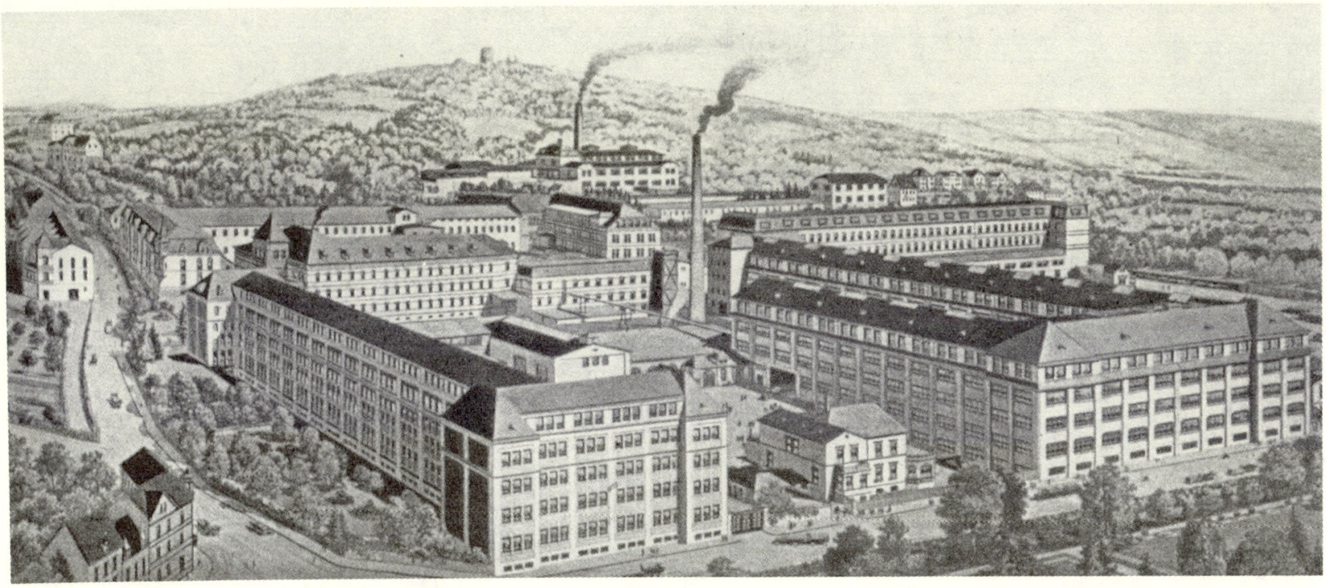

Leitz factory Germany (1926)

made good progress until 1914 when World War I broke out. In order to survive, Leitz began to manufacture and distribute its own laboratory equipment. Its headquarters were located at 30 East 18th Street, near Broadway.

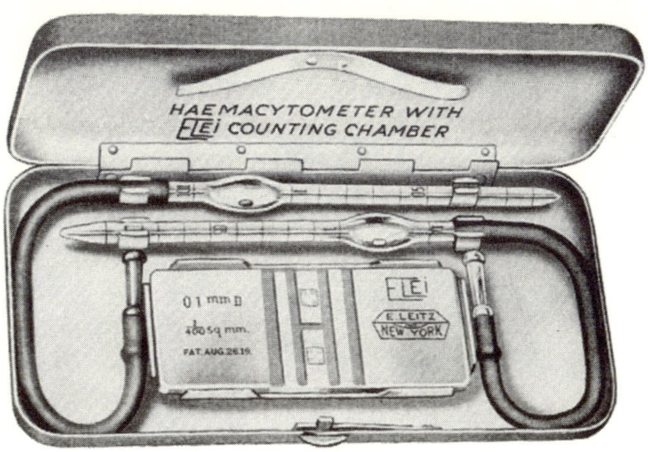

ELei Haemacytometer

In 1917,the Leitz agency was incorporated and since then operated under the name of E. Leitz, Inc. When the United States entered the war, the company was placed as "Alien Property" under the custody of the U.S. Government.

After the end of the war in 1918, importation from Wetzlar resumed and the corporation was purchased from the U.S. Government in 1919 by Alfred Traeger, the same man who had managed it before the outbreak of the war. He was now the sole owner of E. Leitz, Inc., New York. Imports resumed and the corporation needed more space. A five-story building at 60 East 10th Street was rented and refurbished to accommodate both sales and service operations. It remained there until 1937.

In the early 1920's, Leitz U.S.A. flourished till the depression of 1929. In 1910, Leitz catalogues list the following branch offices:

Berlin N.W., 45 Luisenstrasse
Frankfurt/Main, 24 Neue Mainzerstrasse
St. Petersburg, 11 Woskressenski
London, W. 9-15 Oxford Street
New York, 30 East 18th Street
Chicago, 1923 Ogden Avenue.

By 1928, Leitz U.S.A. issued a 290 page catalog listing its entire production program from simple magnifiers to student and laboratory

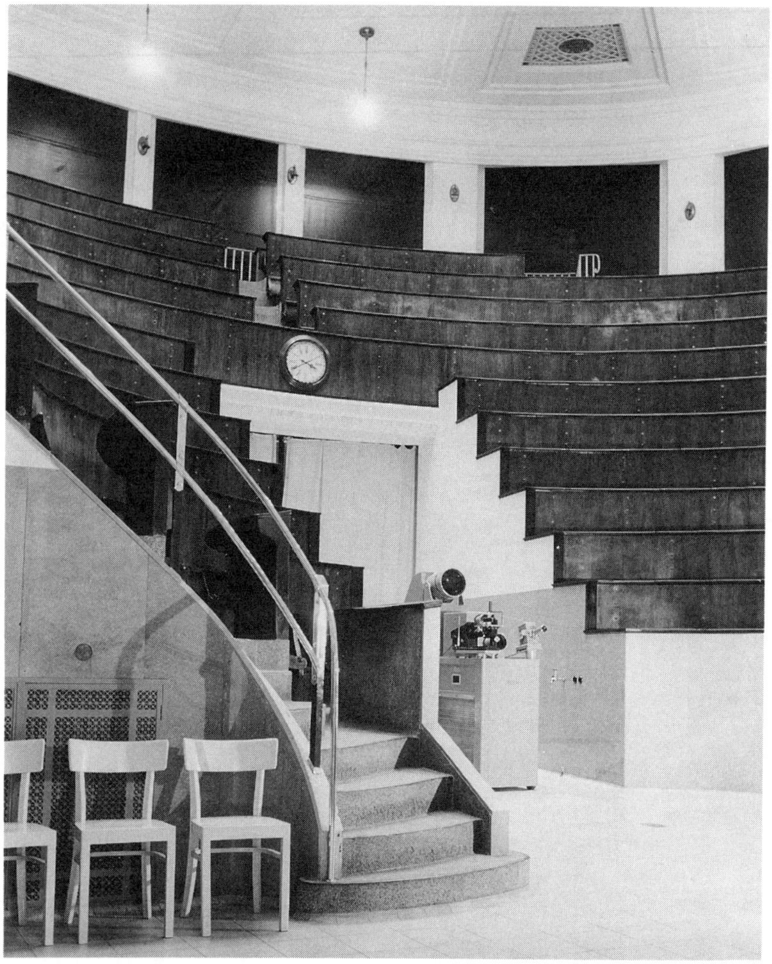

Epidiascope built into the amphitheater

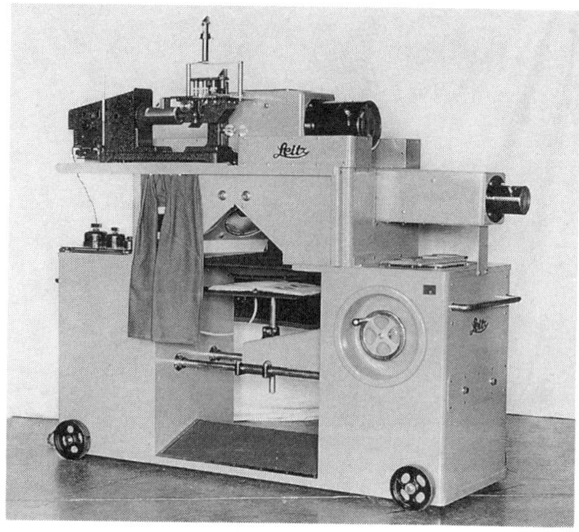

Epidiascope IIIg 800 Watts

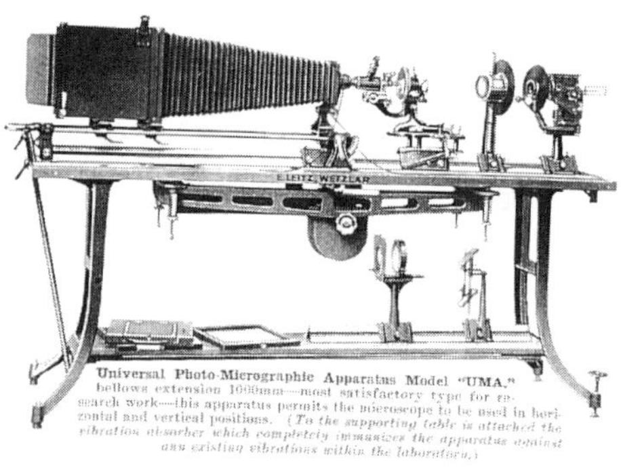

U.M.A. Photomicrographic Apparatus

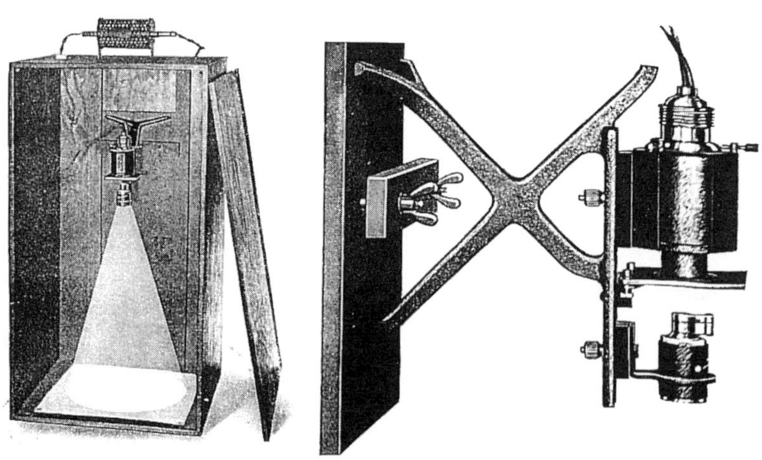

Simple Drawing and Projection Apparatus made in New York, NY, suggested by Prof. Thuringer, University of Oklahoma, Dept. of Histology and Embriology

microscopes, travelling microscopes, comparison microscopes, chemical and petrographical microscopes, objectives and eyepieces for all, metallographs and metallurgical microscopes, stereo binocular instruments, arc lamps, drawing eyepieces, haemacytometers, micro-manipulators (designed in the U.S.), photo-micrographic apparatus, microtomes, colorimeters, microprojectors, opaque projectors, epidiascopes and, last but not least, "Leica", the universal camera.

Quoting from this catalog: "The Leica camera is an exceedingly small but highly developed instrument! ...The Leica is ideal for travelling under all conditions, as it may be carried in any pocket, even the vest pocket. Its reliability under extreme climatic conditions was well demonstrated in its use with the Greenland expedition of the University of Michigan". To quote from a story in "The Pocket Photo Monthly" by Karl A.Barleben F.R.PS: "When the first Leica camera supplies arrived and were shown about, everyone laughed at them. How could such a small thing make pictures. No it was but a brainstorm of the factory and was destined to die out quickly".

Of course, the depression of 1929 almost put another stop to the flourishing business of Leitz instruments in the U.S.A. Its personnel reported to the factory in Wetzlar the latest developments in American Science and Technology. One example was the availability of new 500 Watt incandescent bulbs which were needed for new epidiascopes and lecture hall projectors which were becoming popular in teaching institutions all over the world. The carbon arc lamps, used before then, could be replaced by safer and more convenient sources of light.

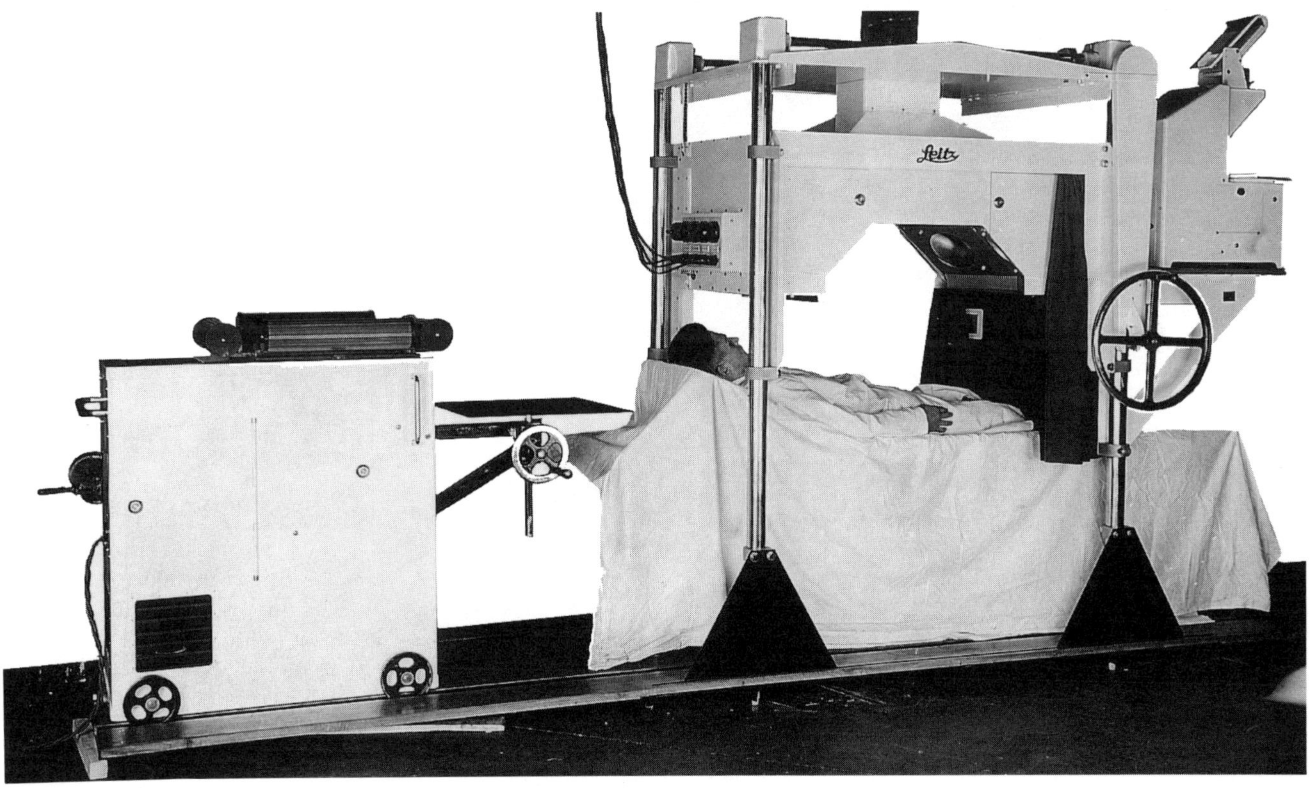

Large Epidiascope for medical schools

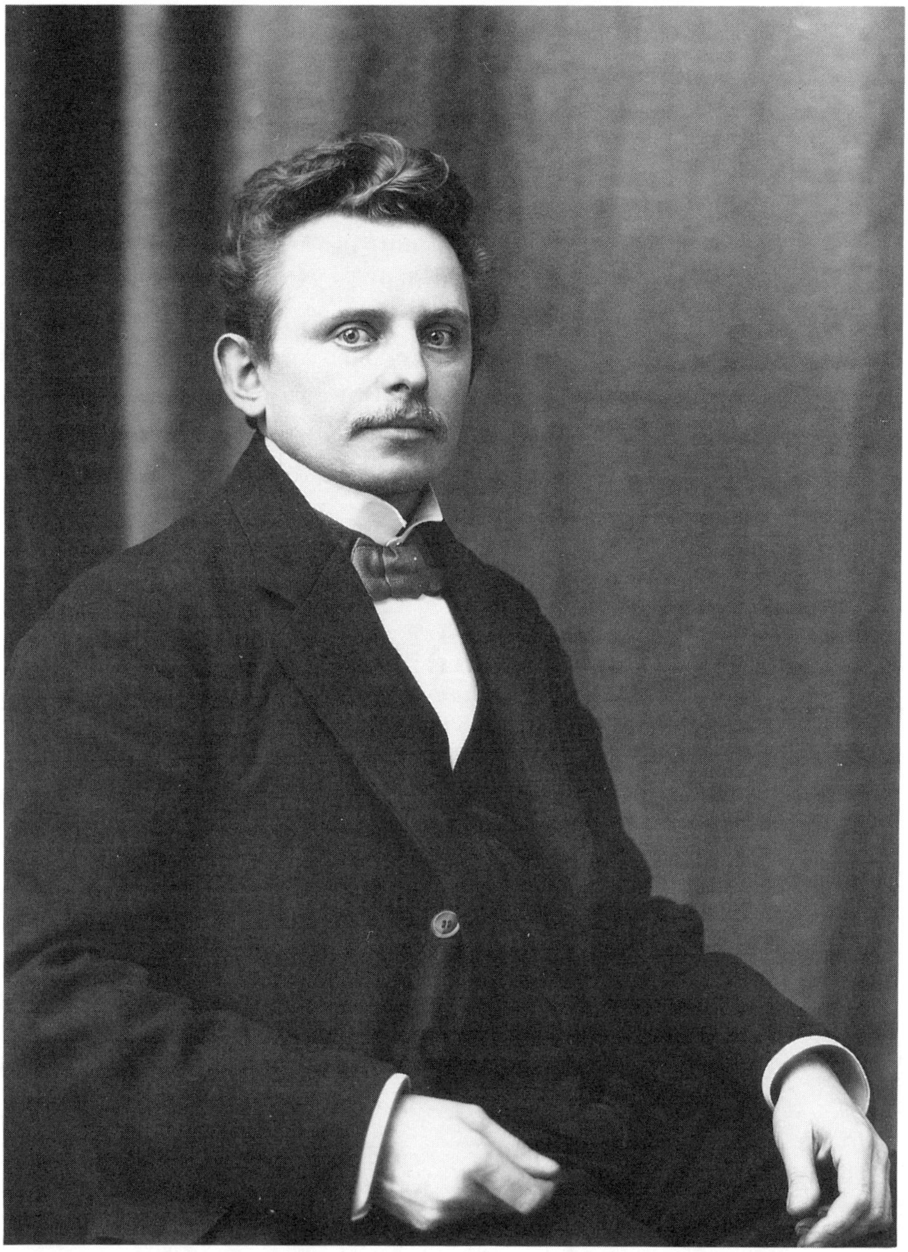

Oskar Barnack as a young man

Chapter 3

HOW BARNACK CAME TO LEITZ

How Barnack came to Leitz

Many variations exist of the story how the Leica was born. The one which follows has the ring of authenticity because it was written by Ernst Leitz III, who related what his grandfather, the founder, had told him about his friend, Oskar Barnack.

In 1910, Leitz Wetzlar was looking for an engineer to take charge of the experimental department for microscopy, where new design ideas are tried and prototypes made. Emil Meschau, who had recently developed a new cine projector with optical compensation, was asked for suggestions. He recommended Oskar Barnack, his friend from their apprentice days at Zeiss Jena, an outstanding mechanic who was rather shy and who did not appear to be able to develop his potential there. Ernst Leitz I wrote to inquire whether Barnack might consider joining Leitz Wetzlar and offered him the position of department head.

Barnack's reply was significant; he wrote: "A business may not consider someone favorably who needs first to acquaint himself with new manufacturing methods and who, as an employee, must take a leave of absence for health reasons, for at least two months a year, not to mention that such a respite at a health spa would for me, as a private citizen, be too expensive!"

Meschau was able to dispel Barnack's reservations and the latter decided to follow Meschau to Wetzlar. Ernst Leitz I soon befriended Barnack, probably because their humanitarian traits were compatible, and saw to it that he had a house to live in that was protected from the fierce East winds, had a glass enclosed porch toward the Southwest, as well as a small vegetable garden.

One day, the new master of the experimentation department began a side project to construct a motion picture camera with which he photographed many traditional events at Wetzlar. In fact, he was a pioneer movie-maker. One problem facing him with the new movie camera was the determination of the correct exposure time. His solution was to begin the design of a small camera for motion picture film with which to go on location, take a few experimental exposures, develop the film and select the proper exposure time. This is how the Ur-Leica was conceived.

The war years (1914 - 1918) and the ruinous inflation which followed, prevented Leitz from producing Barnack's little camera sooner. That decision was made in 1924, but Barnack continued meanwhile to improve and perfect his "exposure device".

As a matter of record, it should be noted here that the first production Leica Models I with serial numbers 100 to 130 were manufactured in 1923 and offered for sale the following year. Leica Models I with serial numbers 131 - 1000 were produced in 1925.

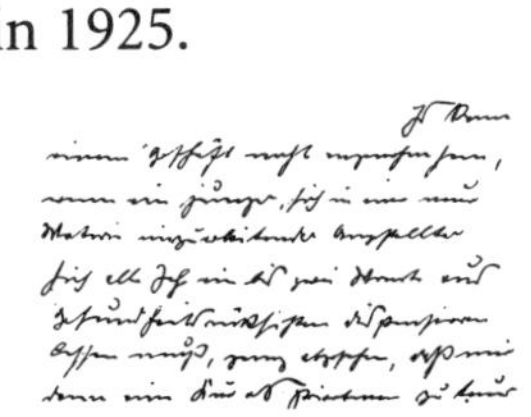

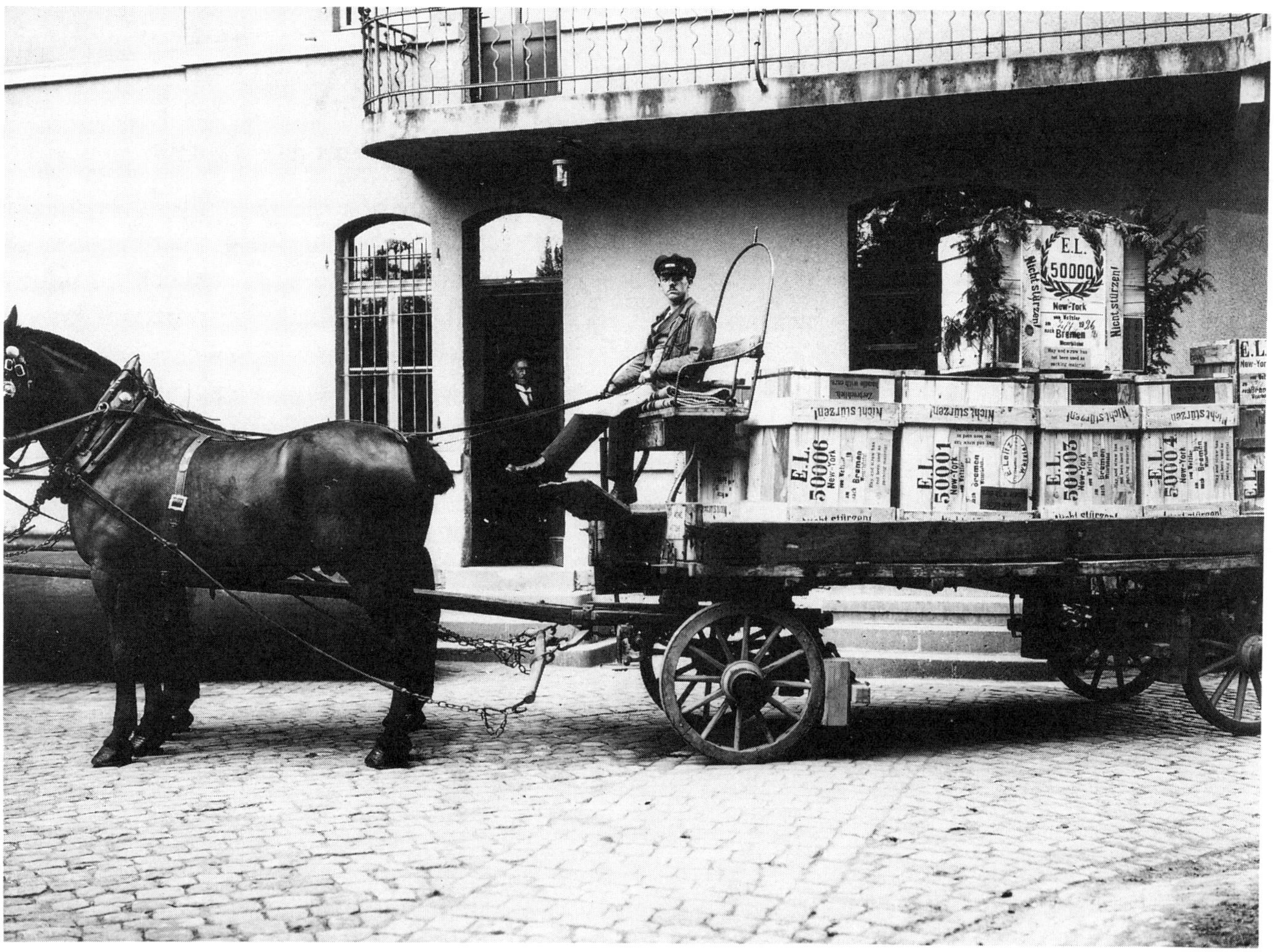

Case No. 50,000 leaves the factory for shipment to New York (1926)

Chapter 4

THE INTRODUCTION OF THE LEICA CAMERA

The Introduction of the Leica Camera, Demanding a Completely New Marketing Approach

Alfred Traeger, until now, concentrated his marketing efforts on the scientific community, namely universities, hospitals, technical research laboratories, medical schools, etc. Dealerships were limited to scientific supply houses whose product knowledge in these highly technical disciplines was limited. The manufacturer and, in the case of Leitz U.S.A., the importer, established branch offices in key markets such as Boston, Detroit, Chicago, St. Louis, San Francisco, and Los Angeles. These offices were staffed by their own personnel, usually trained at the factory, to call upon, sell and explain the purposes and advantages of their instruments to the final consumer.

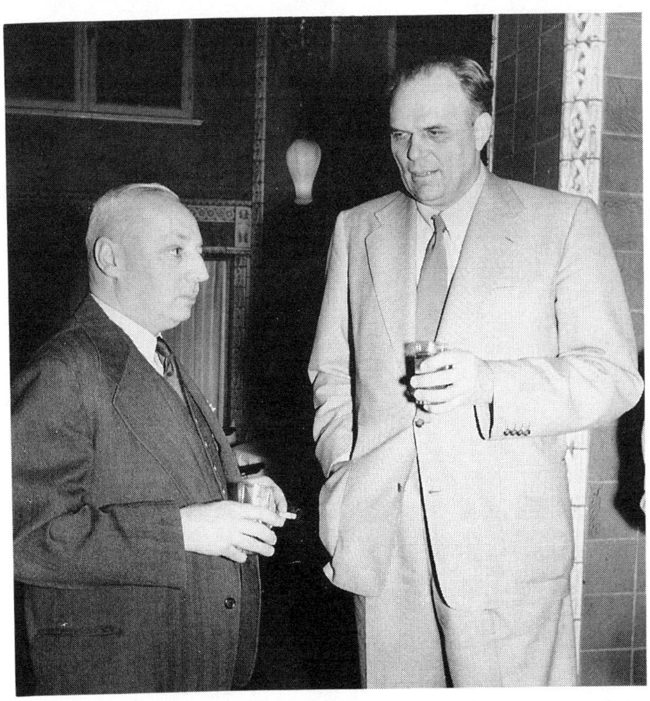

Willard Morgan with Walther Carveth, the Owner and Manager of Leitz in Canada (Toronto)

With the addition of the Leica Camera to the product mix in 1924-25, Leitz U.S.A. had to find talent in the photographic market to introduce and promote their new camera. In 1927, Willard D. Morgan was hired to become sales promotion manager to demonstrate the ease and advantages of using such a small camera, which almost disappeared in the palm of the hand of a man 6' 7" tall. Willard showed fantastic enlargements, projected slides of scenes in California, Arizona and New Mexico. Explorers of the South Pole, China, India, Tibet and Africa took this little camera marvel with them instead of the bulky and heavy 4"x 5" plate cameras, then generally in use. And Leitz N.Y. advertising concentrated on names like Sir Hubert Wilkins, Commander Gatti, L.M. Gould, Dr. Eckner, etc., explorers who were early Leica users.

However, the basic scientific instrument business of the company was not neglected. By 1926, case No. 50,000 was taken to the station for shipment to New York.

In 1931, Ernst Leitz III came to New York. He stayed several months, and joined a rowing club on the East River, but his prime mission was to analyze the U.S. market potential in all phases of production, with emphasis on Leica sales.

The microscope business in the late 20s and early 30s was strong in spite of competition from American Optical (Spencer Lens Co.), Buffalo, NY and Bausch & Lomb, Rochester, NY. Medical

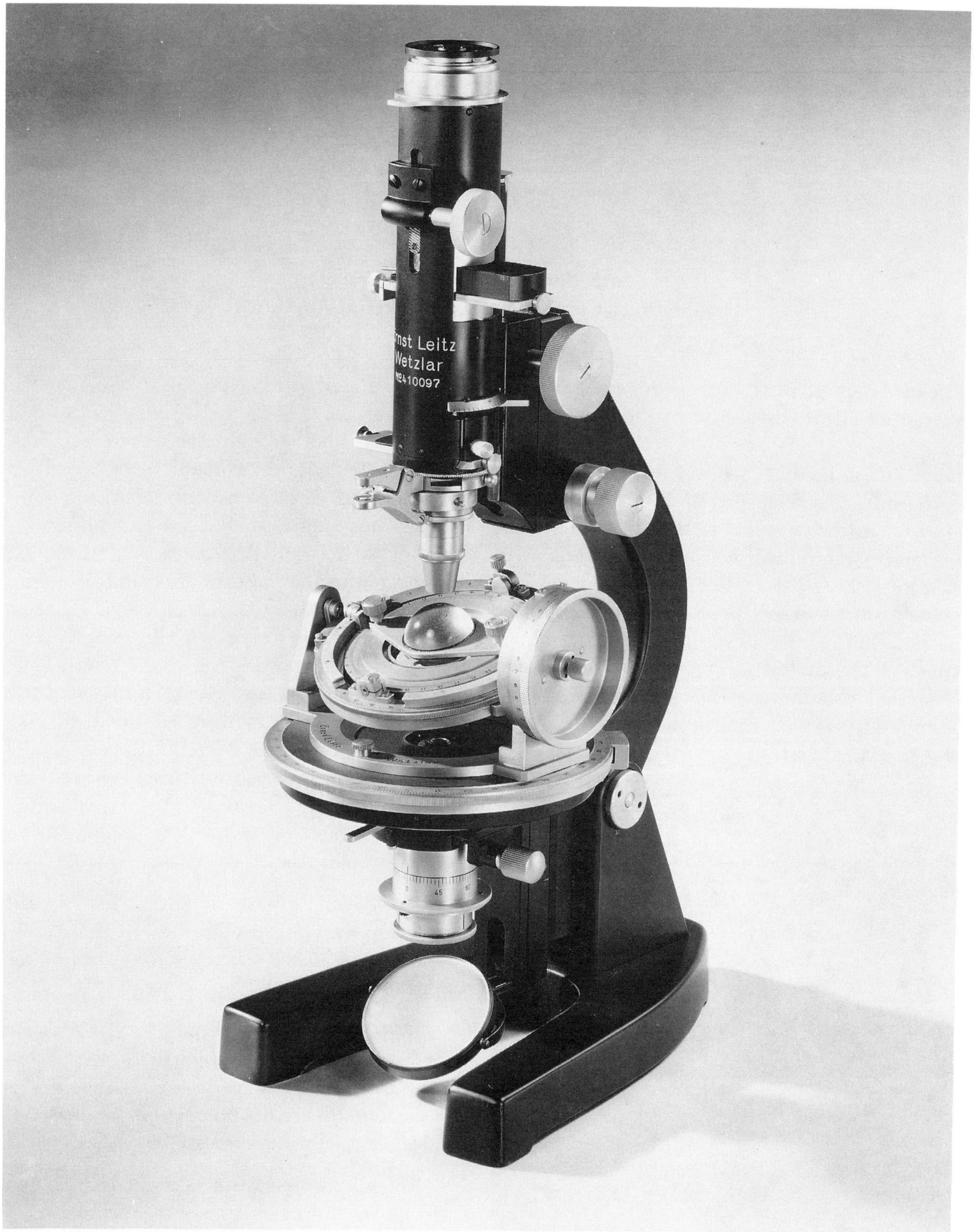

CM Microscope with Federow Universal Stage designed by Max Berek

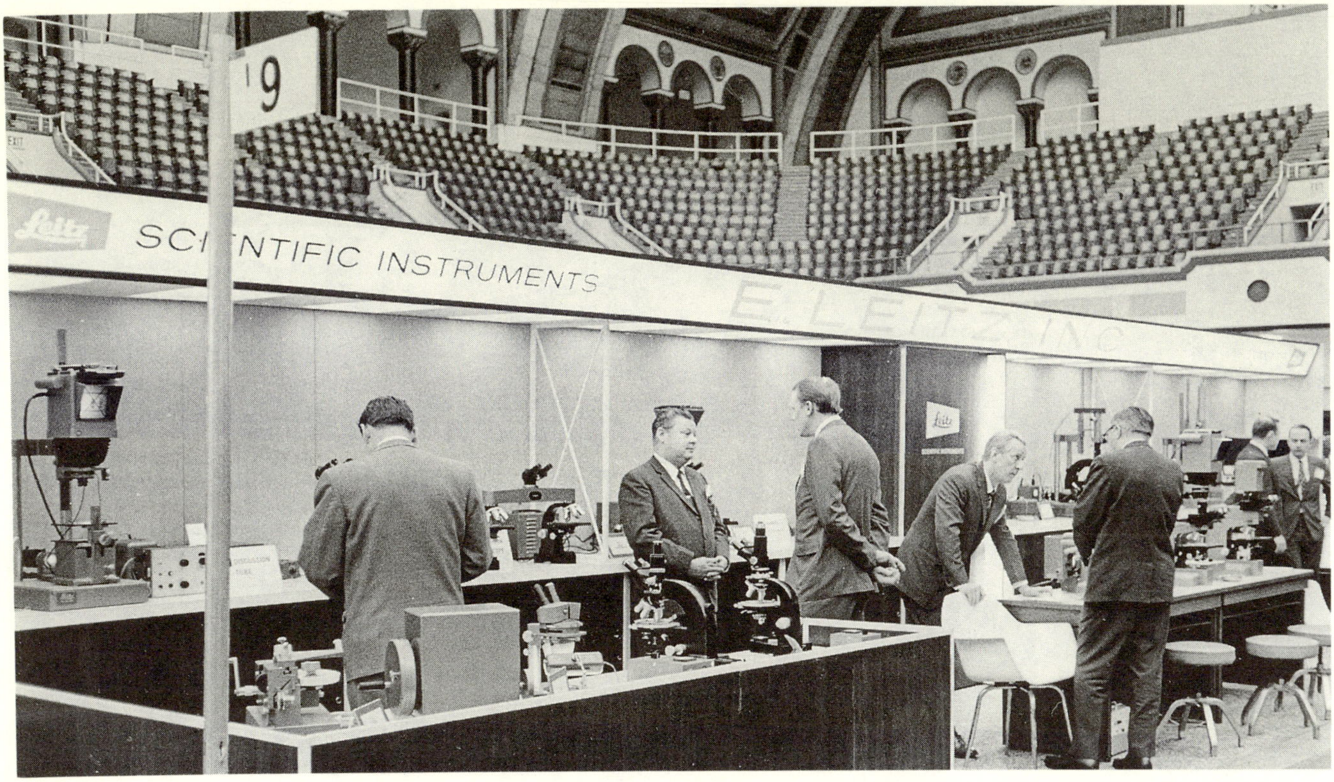

E. Leitz, Inc. Exhibit Booth for Scientific Instruments (Atlantic City)

students in the U.S. were required to purchase their own microscopes and Leitz U.S.A. succeeded in gaining and maintaining a considerable market share. Many hundreds of monocular (LS-10 $ 150.00) and binocular (BST $ 255.00) Leitz microscopes were sold, often just outside the portals of the medical schools themselves by manufacturer's (importer's) representatives, through College Book Stores, or through local laboratory supply dealers. The Japanese did not yet participate in this business, nor did Zeiss Jena.

Student microscopes, however, were not the most important items. New developments in research, both biological and industrial, demanded sophisticated instruments such as the "Panphot" universal photomicrographic apparatus. This new research tool combined a large research instrument for both transmitted and reflected light with a 4" x 5" camera. The design engineer in Wetzlar was Mr. Riepert.

During this time metallurgical instrumentation became increasingly sophisticated. The large Micro-Metallograph with integrated camera, the MM-5, was adopted in the U.S.A. by most of the important industrial laboratories. Microtomes for rapid tissue diagnoses were sold to hospitals nationwide and the geology departments of many universities purchased petrographic microscopes perfected in Wetzlar by Prof. Dr. Max Berek.

In 1937, Leitz engineers Maisch and Heine, conceived a new microscope to be called "Ortholux". The design deviated radically from the conventional type research microscope. It was turned away from the observer and had the light source built-in, fully aligned. Base and arm were now one solid piece, imparting sturdiness. Focusing the image was done by raising or lowering the mechanical stage. The instrument could be used for transmitted as well as reflected light. There is no doubt that this ingenious design revolutionized the shape of microscopes to come.

As a tool for early cancer diagnosis in 1934, Emmerich von Hamm, M.D., at Louisiana State

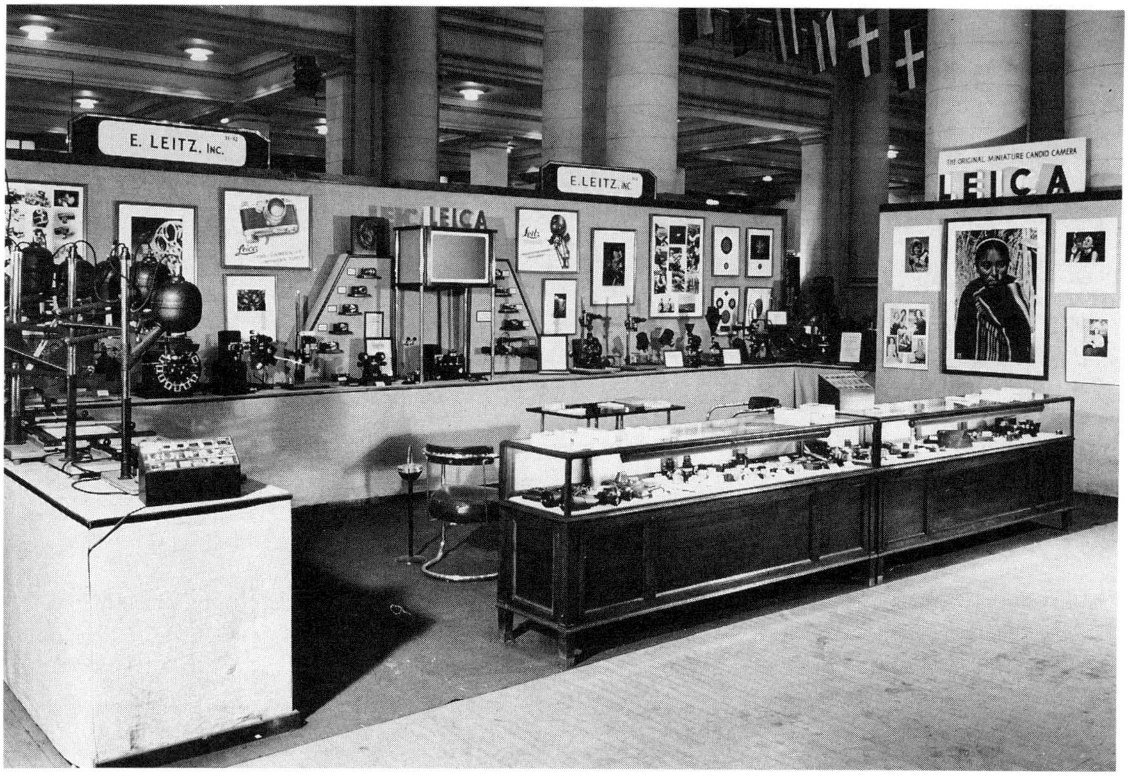

Leica Exhibit booth before World War II

Panphot Photomicrographic Apparatus

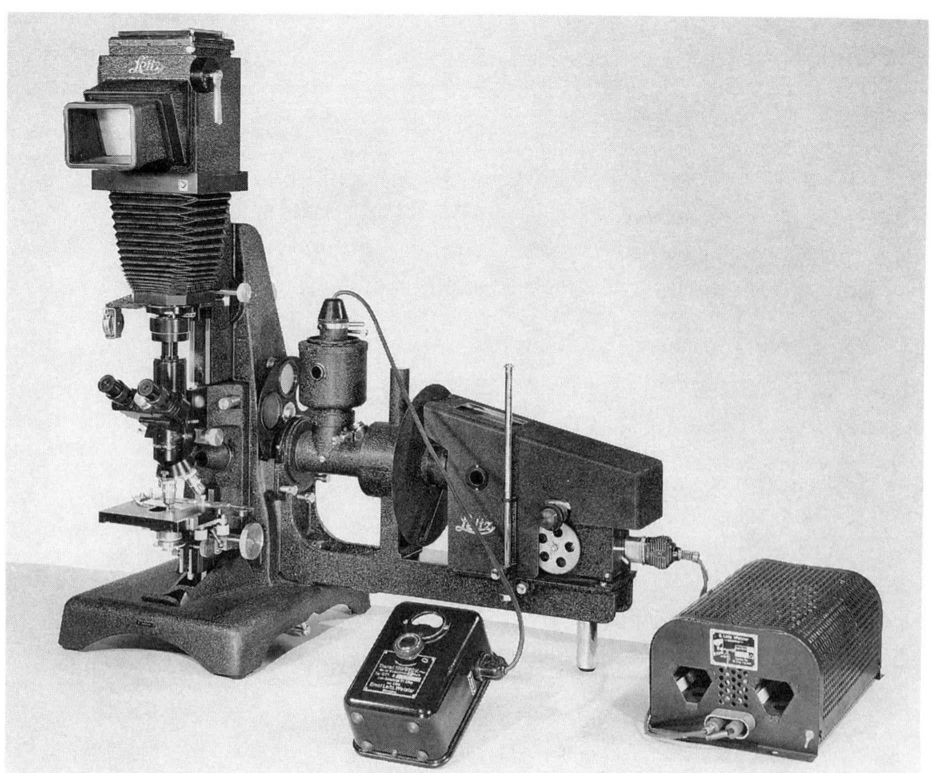

University and Charity Hospital in New Orleans, used Heine's newly invented "Ultropak" to examine a patient's biopsy. This same tool was also successfully used to examine welding seams in metal construction. Heine was head of the instrument design department at Leitz Wetzlar.

During this time the design of a new toolmakers microscope, featuring an "intermediate image" was developed, perhaps due to the need for large numbers of precision parts for the Leica camera. This "image" is the replica of the object to be inspected, together with an image of the master part with which the first part is to be compared. Rather than sell such equipment through Leitz New York, it was imported and distributed by a specialist in the toolmakers field, the George Scherr Company, 128 Lafayette Street, New York.

It now became a challenge to develop market potential for the new miniature camera.Willard Morgan and his brother Gilbert in Hollywood, owner of the Morgan Camera Shop, started to meet this challenge. Leitz U.S.A. effectively used its connections to the scientific community to promote the camera. However, this was not enough, as Ernst Leitz III observed during his U.S. visit in 1931. New dealerships had to be organized, photo salesmen were needed and had to be trained, and the dealers had to be persuaded that this new camera with a film area of just over the size of a postage stamp had a future. The American public was still accustomed to cameras with the negative size of 4"x 5". It was hard to believe that a film area of l" x l.5" could produce similar results. Of course, the films (35 mm) of that time were grainy and slow and Leitz New York was forced to establish their own "Fine Grain Laboratory".

Soon dealerships were established in the important business centers of the country and the Leica began to make a name for itself. Again a depression (1929) slowed the progress in the advancement of Leica sales.

By 1931, new economic improvement became evident. Morgan had started a new magazine called

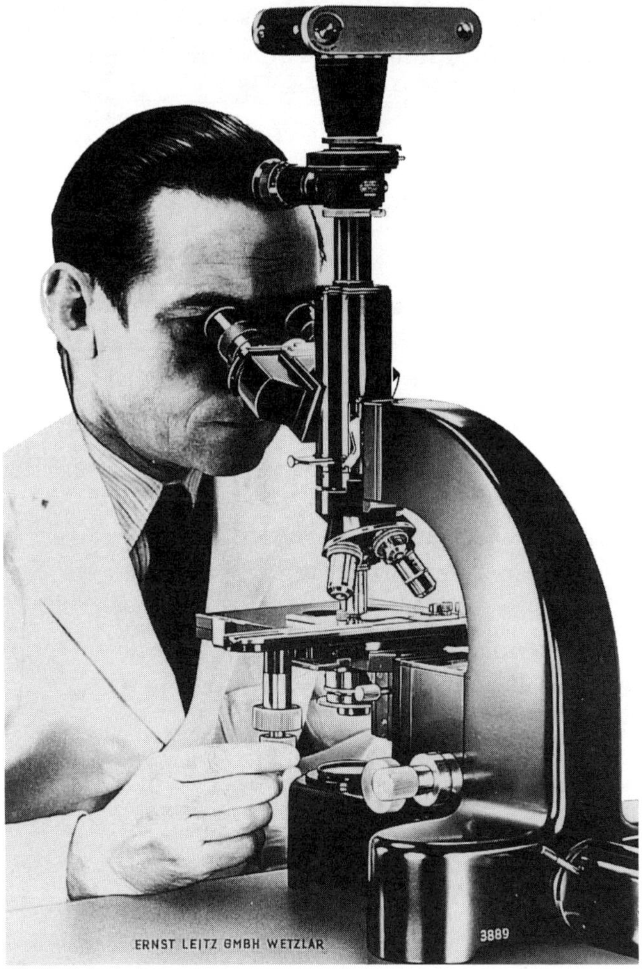

Research Microscope "Ortholux" 1935

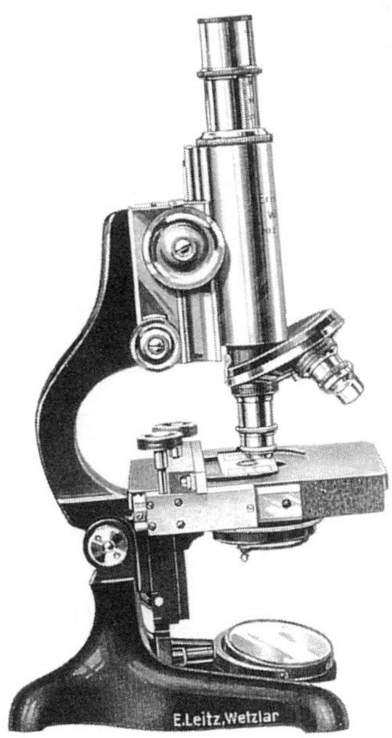

Medical Student Microscope

"Leica Photography", together with Karl Barleben. Pictures and practical hints on how to take them became the chief message of the new journal. It was distributed to all registered owners of miniature cameras. New lenses, when added to production, were shown and discussed.

In the same year, Leitz Wetzlar repurchased 50% of the assets of Leitz New York from Alfred Traeger, who suffered from crippling arthritis. Four years later, in 1935, Ernst Leitz III again came to New York to successfully negotiate with Traeger for the balance of the assets of Leitz New York. Once more in the hands of the Wetzlar factory, Leitz New York needed a new management team. Wolfgang Zieler became President and Alfred Boch Vice President/Treasurer. Zieler was a friend of Ernst Leitz III dating back to their days as pupils in a private boarding school in Wickersdorff , near Saalfeld.

Alfred Boch's background is of interest, because, as a mechanic's apprentice he actually worked in the workshop supervised by Oskar Barnack, where the first Leica Cameras were produced. Alfred was born in 1904 in Wilno, then a part of Russia. From 1914 to 1918, he and his family were interned in Siberia where he learned to read and write Russian fluently. In 1918, he returned to Wilno but when the Russian/Polish War (1919) broke out, he and his brother Karl fled through Lithuania by horse cart and on foot to reach the German border. The Boch brothers ultimately reached Wetzlar where they had relatives. Ernst Leitz I hired Alfred as a mechanic's apprentice and this is how his career with Leitz began.

Leica promoter, Willard Morgan, in 1935 decided to leave Leitz New York and, together with Henry Lester, published his first Leica Manual. The following year he joined the staff of Life Magazine, a new weekly newspaper that featured photo journalism by outstanding photographers such as Alfred Eisenstaedt and others. 35 mm photography was by now fully accepted. The possibilities were endless and the designers in Wetzlar kept offering a flow of accessories, lenses, enlargers, copying and close-up devices which gave the photographic public enormous new potential for expression. In fact, photo journals such as Life, brought public awareness close to the events of the day in politics and in the society as a whole.

The new management at Leitz New York took advantage of this trend and promoted their camera by featuring exhibitions of outstanding Leica photographs in all parts of the country. Similarly, Leica lectures and exhibits were held at local camera clubs and in 1937 Anton Baumann was brought to the U.S.A. for the purpose of demonstrating the diversity of the growing Leitz-Leica System of Photography.

As early as 1934, the Morgan Camera News, an internal newsletter was published by the Morgan Camera Shop in Hollywood, and distributed to all his customers. Many celebrities were early Leica users. Morgan counselled them on how to take pictures at Santa Anita horse races. The great bridge in San Francisco was under construction and Mr. Daves, author of Warner Brothers latest films, was using his Leica shots as the authentic production backdrops. After studying and writing about the Arizona Indians, the Baroness von Langenn, a Berlin correspondent of a European Newspaper, illustrated all her articles with Leica photographs, developed and printed by Morgan. Even Ernst Udet, the German World War I flying ace, was one of his customers. Joan Blondell kept a Leica record of her new born son. Frank Capra, producer of the four star pictures for Columbia Studios, was a Leica fan as early as 1935.

The Saturday Evening Post as well as Collier's magazine and Time were by now featuring photographs taken with the Leica. Thomas McAvoy had taken a series of photos of President Franklin D. Roosevelt at his desk. With the expression of the President constantly changing, he was a most interesting man to picture with the rapid firing Leica.

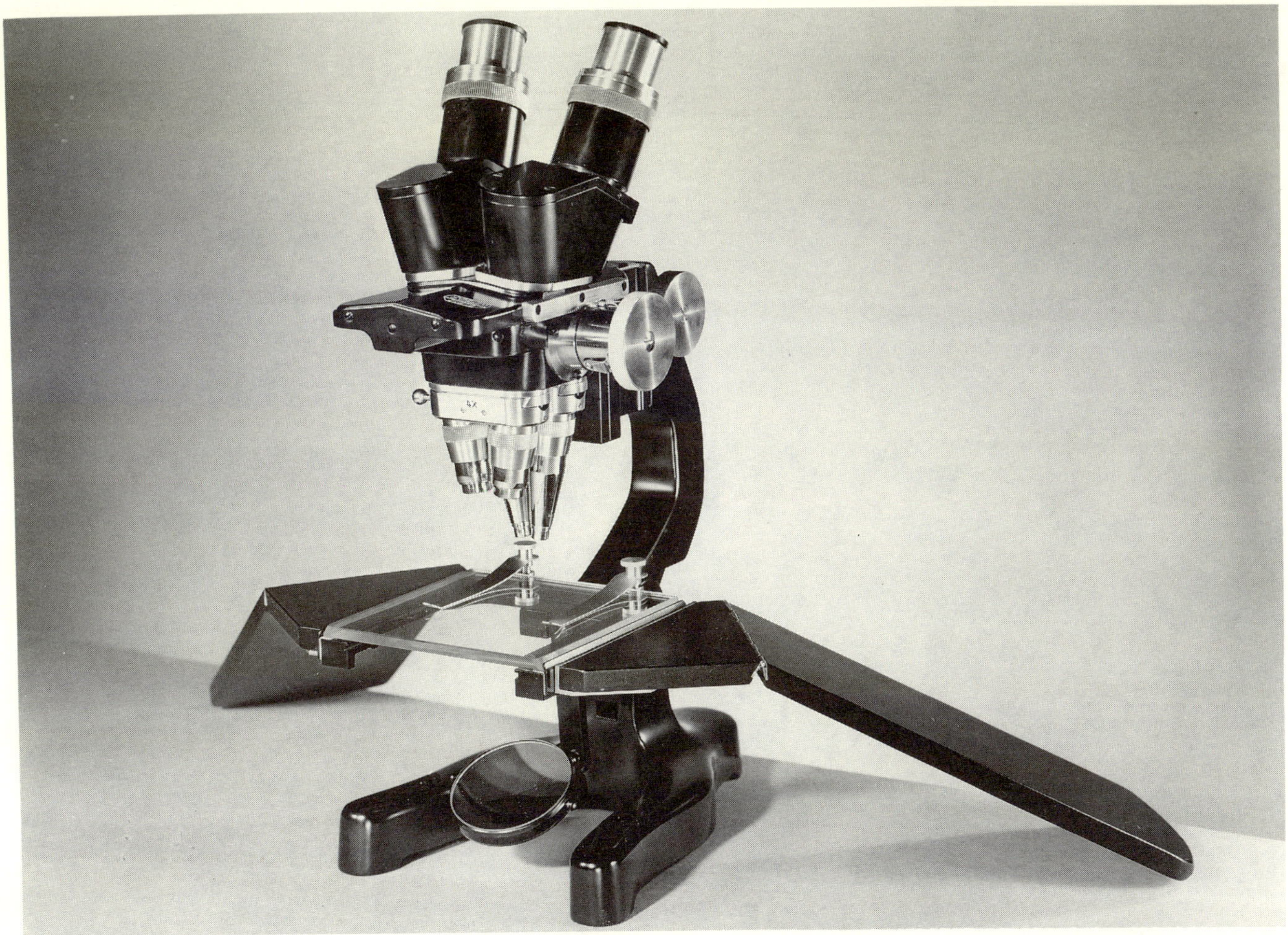

Stereoscopic Binocular Microscope after "Greenough" by Leitz

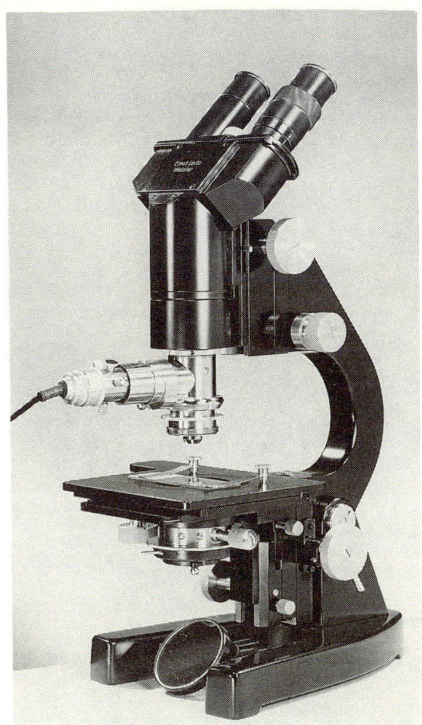

UBS Microscope
with Ultropak Illuminator

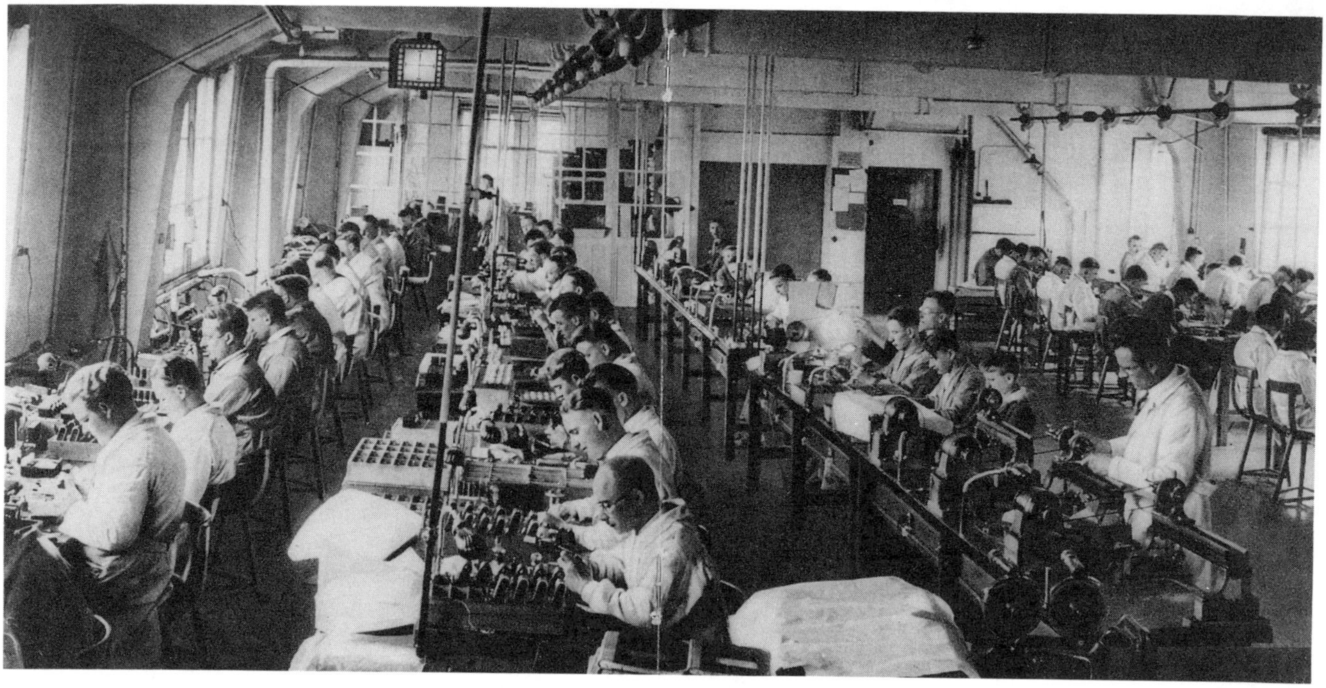

Leica camera assembly in Wetzlar (1932)

Chapter 5

LEITZ NEW YORK UNDER NEW MANAGEMENT

Leitz New York Under New Management
by Zieler, Boch & Wechsler

As indicated in the previous chapter, the Leica Camera had by now grown from a novelty to a completely new system of photography. By 1932, the camera contained a built-in rangefinder as well as means for coupling a family of seven interchangeable lenses. In 1935, the Leica offered a top shutter speed of 1/1000th second, a rapid film/shutter winding mechanism as well as slow speeds between 1 second and 1/20th second. Leica photography was by now accepted worldwide.

Competition by other camera makers was not far behind. Zeiss had introduced the Contax camera; Kodak Nagel its own 35 mm camera; as did Argus, to name a few. Most importantly, however, Leopold D. Mannes and Les Godowsky introduced 35 mm Kodachrome color film. More than 175,000 Leicas had been sold by now.

The new management had the task of expanding the U.S. market, both with respect to Leitz Scientific Instruments as well as the Leica Camera. Alfred Boch concentrated on the latter while Zieler continued to develop the instrument business. Alfred Traeger came to visit once or twice a week while Wechsler dealt with the affairs of expanding personnel. One of the major immediate decisions involved the move to a larger and more prestigious location. 730 Fifth Avenue was chosen in 1936, across from the new headquarters of Tiffany & Co., then under construction.

To continue the success of the Leica Photography Magazine, two new employees were hired, namely George Hesse and Gus Wolfmann, both with backgrounds in photography and journalism. New sales territories were chosen, branch offices enlarged, and personnel hired to deal with the expanding markets.

The U.S.A. had become the largest market for Leitz Scientific Instruments and the Leica Camera.

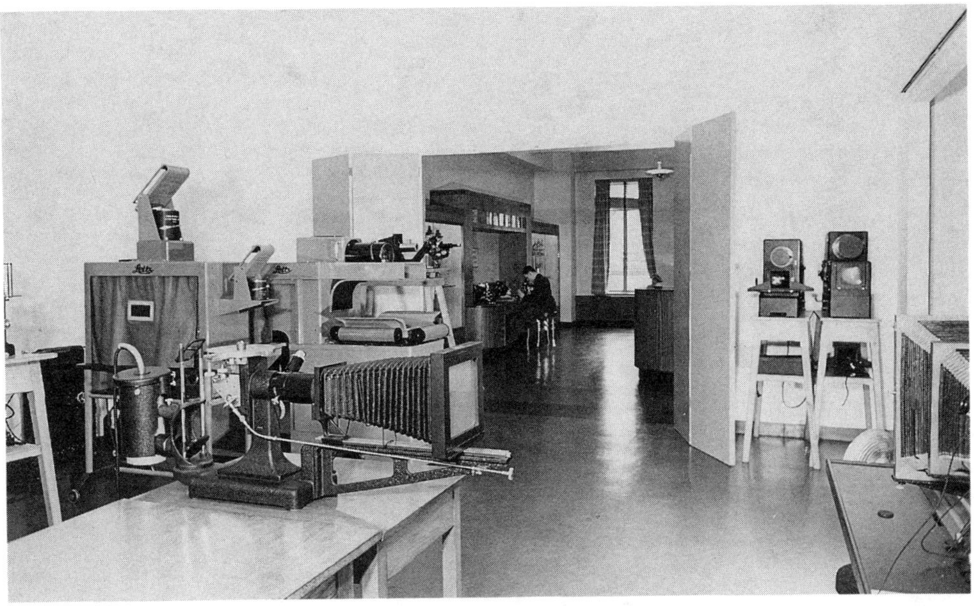

Part of the showroom at 730 Fifth Avenue (1938)

University laboratories, medical schools, industrial manufacturing plants and independent research laboratories were the principle customers for their scientific instruments.

Meanwhile the general public adopted the Leica as their choice for photography. This was accomplished by sending specially trained experts throughout the U.S.A. on International Leica Exhibits. They demonstrated the camera and its accessories before camera clubs with slide shows and enlargements taken by outstanding photographers. Local contests were awarded with the Barnack Medal as first prize. Then, Anton Baumann was brought from the factory to lecture on photo techniques, including producing large format enlargements. His greatest asset at the time was the availability of Kodachrome Film in 1936-37, enabling him to project tremendous screen images in living color to spell-bound audiences. He also published a Leica Color book.

Baumann's assistant, Julius Huisgen, who had been a member of Heinrich Stoeckler's Leica School in Germany, came to the U.S.A. in 1938. They travelled together on occasion and Julius maintained Baumann's darkroom facilities, making enlargements and duplicating slide shows. Baumann's fatal accident in 1939 left a vacuum. Huisgen tried his best to fill it and became a most valuable member of the company, acting as a teacher, photographer and ultimately sales representative with a territory including Washington, D.C.

Financial matters began to interfere. Particularly, the currency exchange between the U.S. Dollar vs. the German Mark was increased, causing the cost of merchandise from the factory to rise by 50%. This change threatened to curb the Leitz U.S.A. business entirely; however, a unique solution was in sight. It was the so-called Aski-Mark. E. Leitz, Inc. New York, purchased American cotton and sold it to a German importer who, in turn, used Aski-marks (at the old rate of exchange) to repay Leitz New York. This method lasted till 1939 when, after the outbreak of World War II, new and more difficult problems arose. A naval blockade stopped all commercial shipping and other ways had to be found to keep Leitz New York afloat.

As was the case during World War I, Leitz New York had the alternative to manufacture products that fit into its marketing mix locally. Repair facilities were now enlarged to include manufacturing and the technicians produced a photo electric colorimeter. This product was suitable for hospital use, with the capability to determine up to 40 clinical tests such as phosphates, hemoglobulin, uric acid, urea nitrogen, etc. Microscope lamps were made, but the main thrust of production centered around sliding focusing copying attachments for the Leica, slide and film viewers, universal finders, flash units, projectors, enlargers, filters, and carrying cases. Manufacturing support in terms of design drawings, was still available from Wetzlar.

When the U.S.A. entered the war in 1941, the Government again placed the firm under the custody of Alien Property. Under that administration, its skilled staff of mechanics and optical specialists became engaged in producing precision-optical war material, such as optical components for gun sites, etc. The company was now closely monitored and an FBI agent was permanently assigned. John Slacks was President and supervised all activities.

In 1941, the U.S. Navy Department in Washington was the first to experiment with 35 mm cameras to photograph a fluorescent screen. The lens employed to copy the screen was an F/1-5 Xenon 50 mm together with a 250 exposure Leica camera. The resulting films were inspected by means of a Leitz New York desk viewer. U.S.manufactured Sliding Focusing Copying attachments were now available in various models as were synchronized flash units models V, Va and VI for the different Leica models.

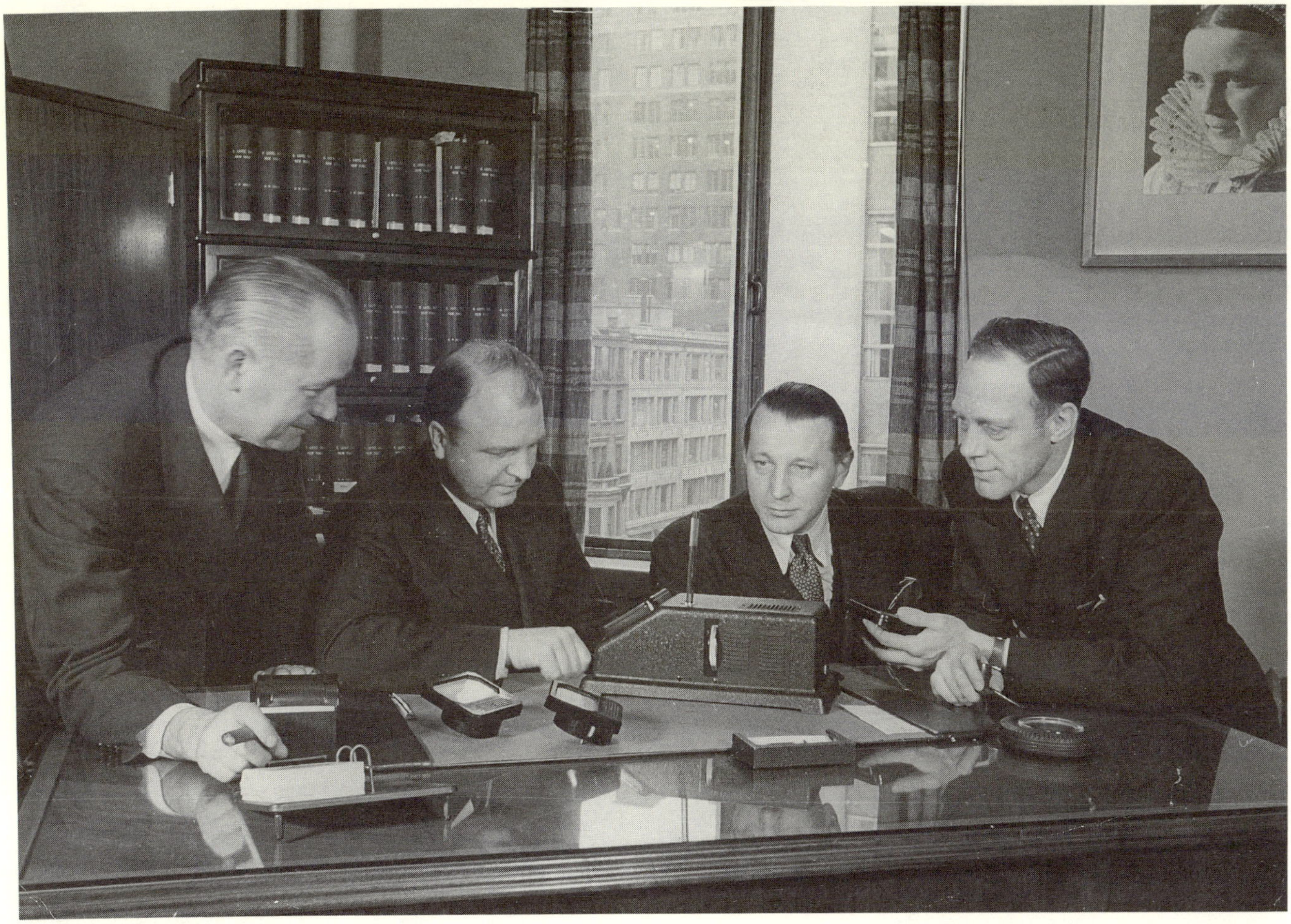

Prototype Colorimeter
From left to right: H. Wechsler, V.P.,
Wolfgang Zieler, President, Alfred Boch, V.P.,
Chick Kidner. A.P.S.

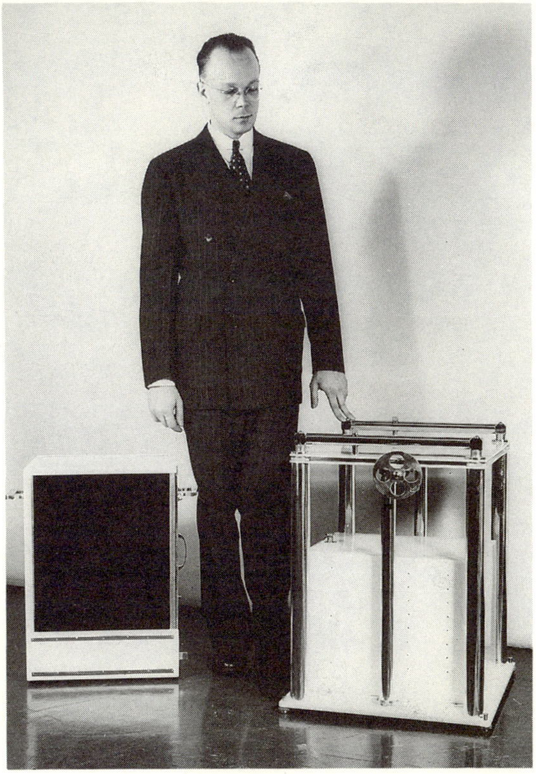

George Hesse with U.S. Navy Fluorographic Unit

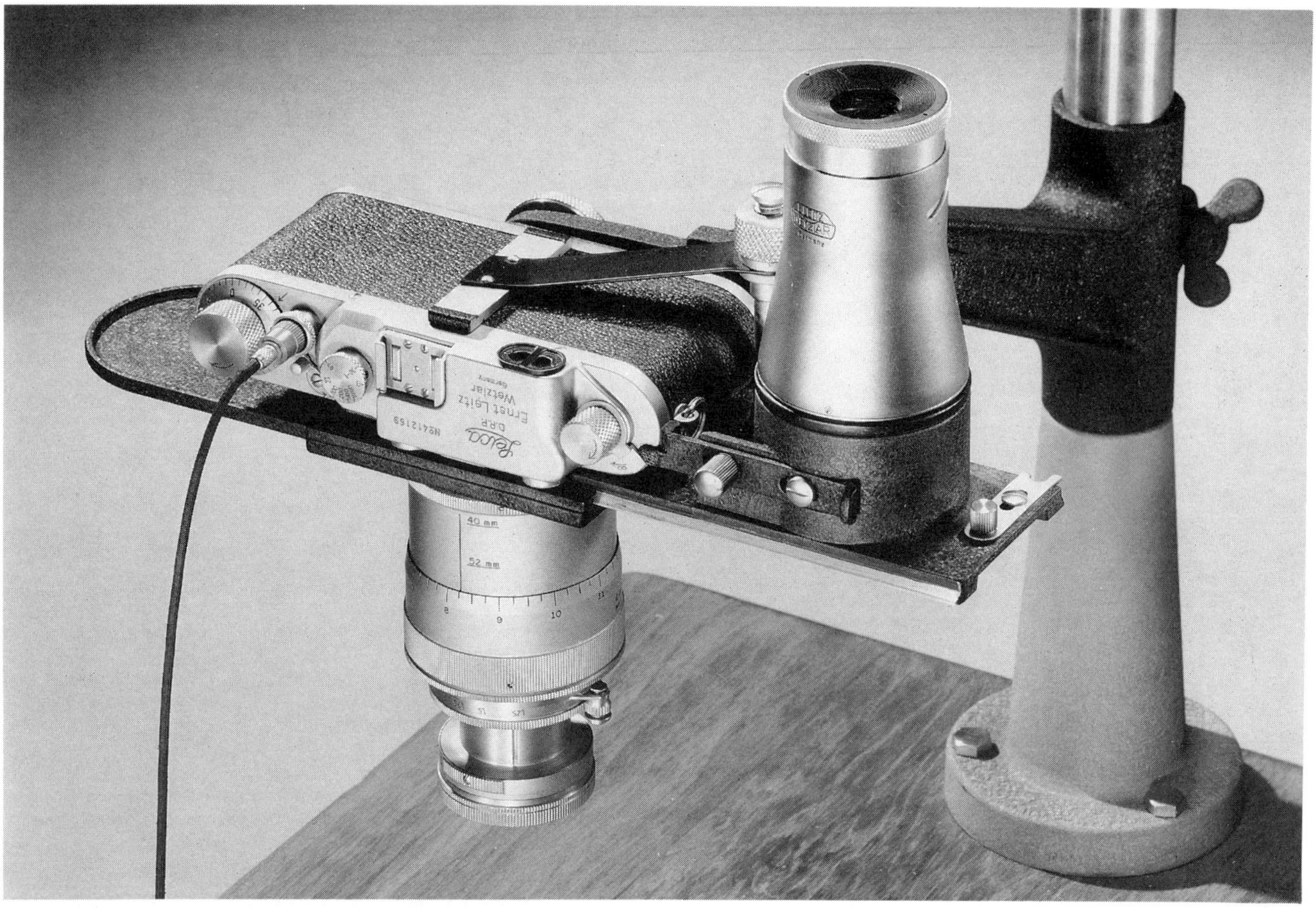

Copy stand and focusing mount made by Leitz New York

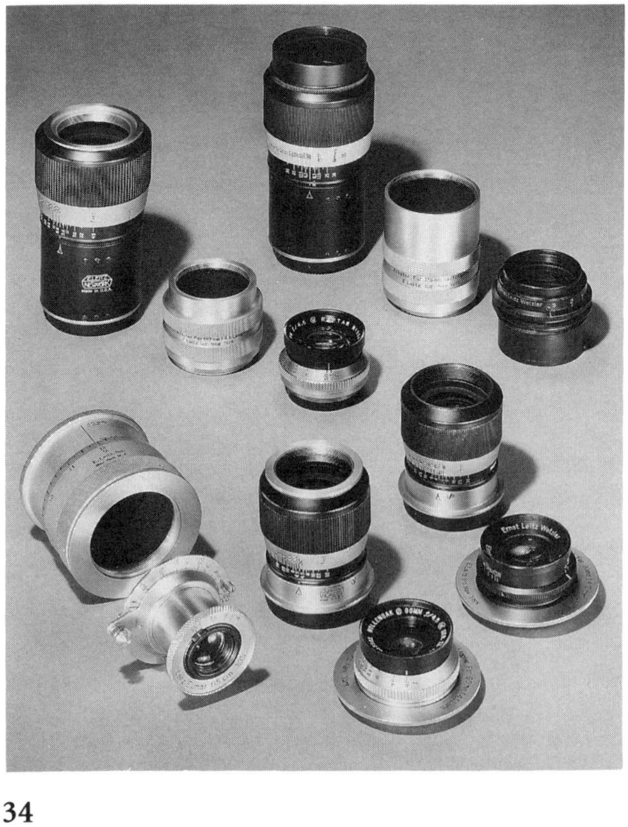

Focusing mounts for various interchangeable lenses made by Leitz New York

For the record, here are some prices for the latest model Leica Cameras, effective March 15, 1941, as follows:

	U.S.$
LOOEH Leica Model III, black, with Summitar 50 mm F:2 lens	253.50
LOODU Leica Model IIIB, chrome, with Summitar 50 mm F:2 lens	273.00
Weston Leicameter	24.00
WOOSU Image reversing angle Viewfinder	19.50
VUDER Leitz Desk Viewer with 3 element achromatic lens	48.00

It deserves to be mentioned that by 1939 the Signal Corps of the U.S. Army already purchased a number of Leica outfits for its observers on the European front. Photographers with such equipment were less likely to suffer casualties when compared with those who carried bulky 4" x 5" Graflex cameras. At that time, the Army already was interested in purchasing a larger quantity. Leitz New York could not meet this order, except for a small quantity which was bought back from registered Leica owners, refurbished, and then made available to the military.

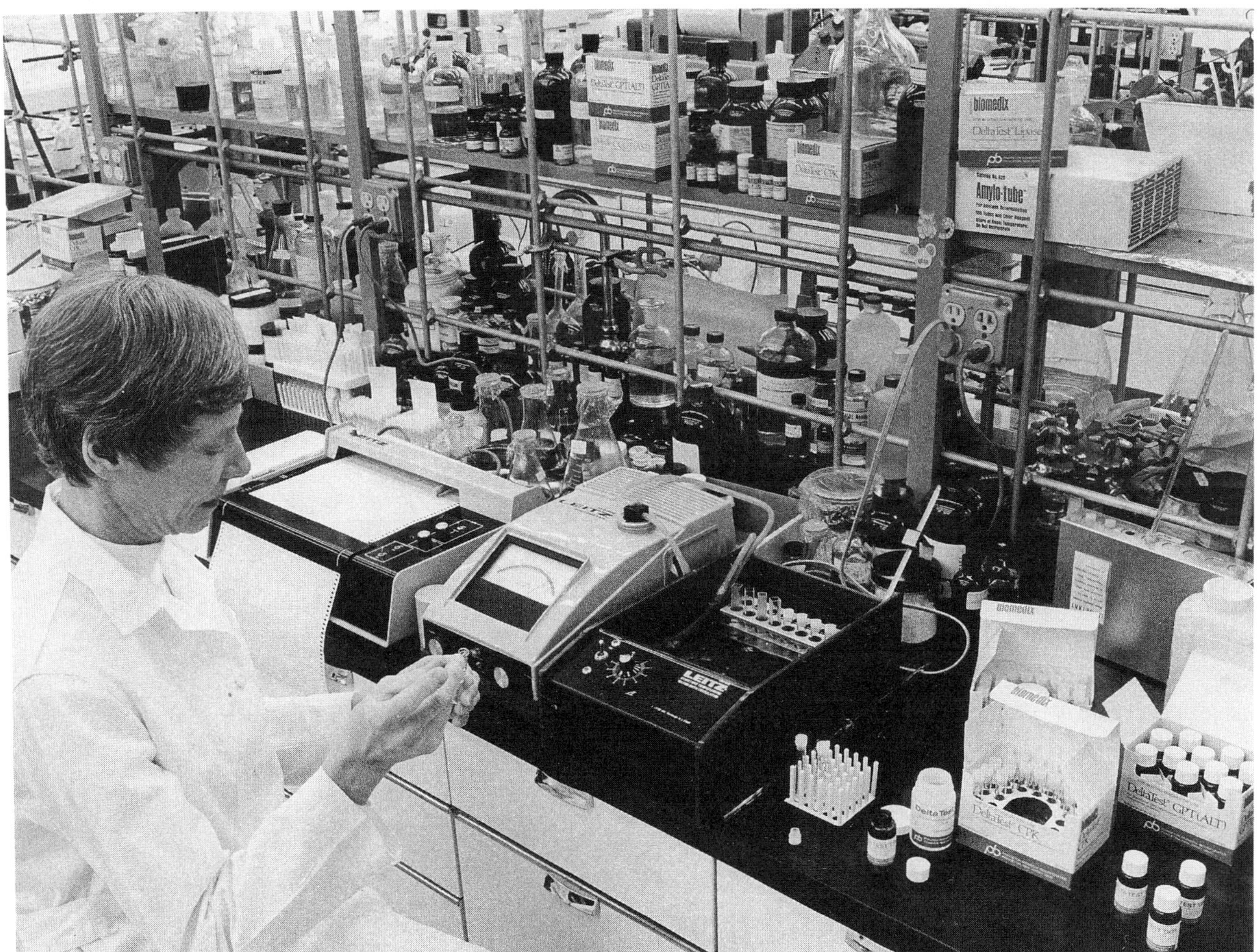

Leitz New York Photoelectric Colorimeter in a clinical laboratory setting.

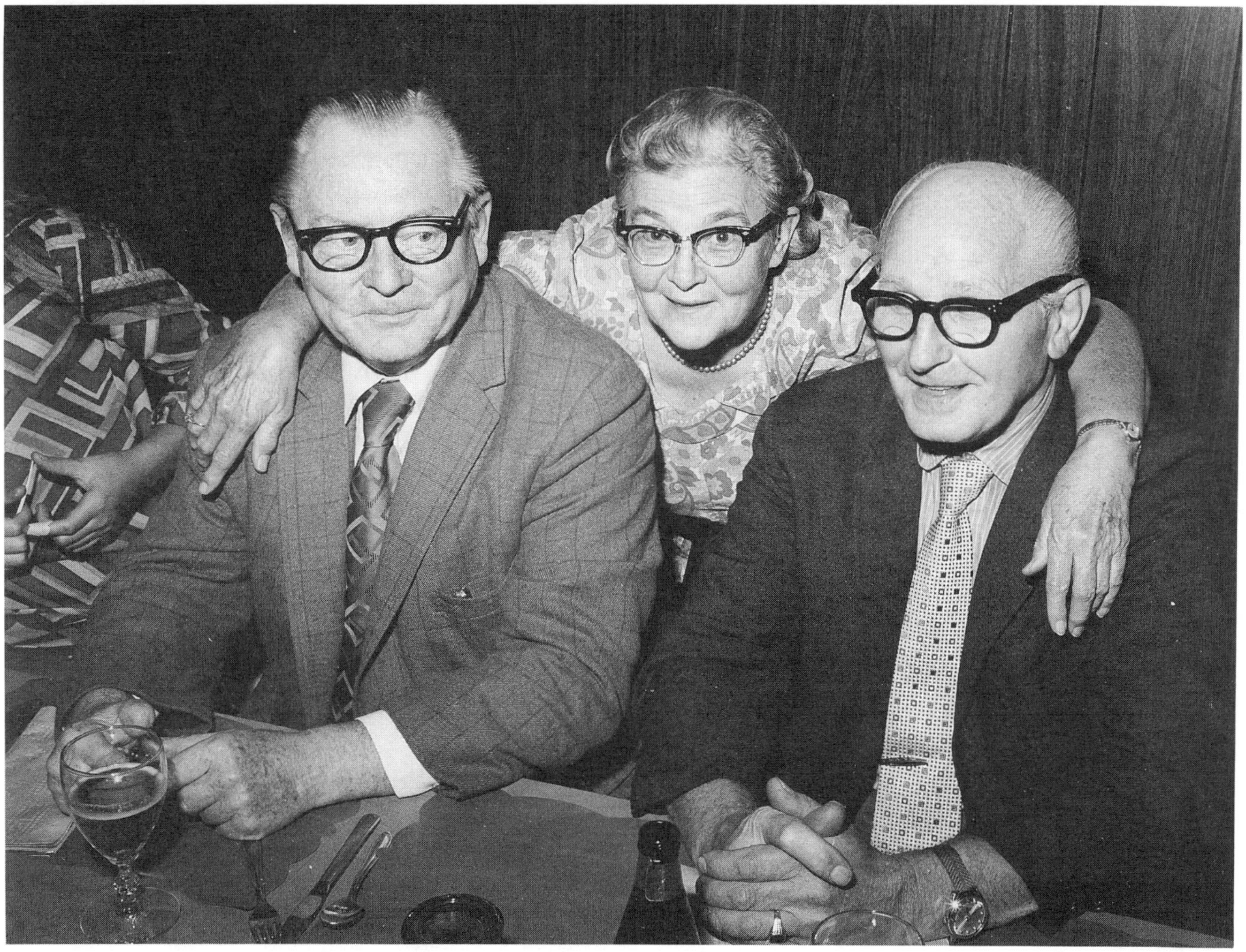

Margaret Vetter with two of the most important mechanics at Leitz New York during World War II, namely: Willi Ulm and Gus Borchert

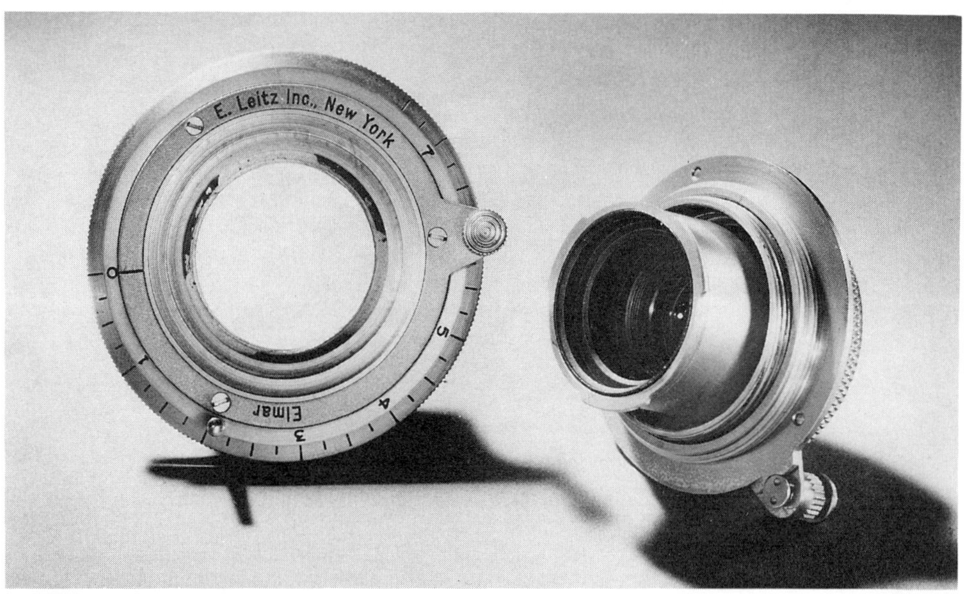

Focusing mount for Leica lenses made by Leitz New York

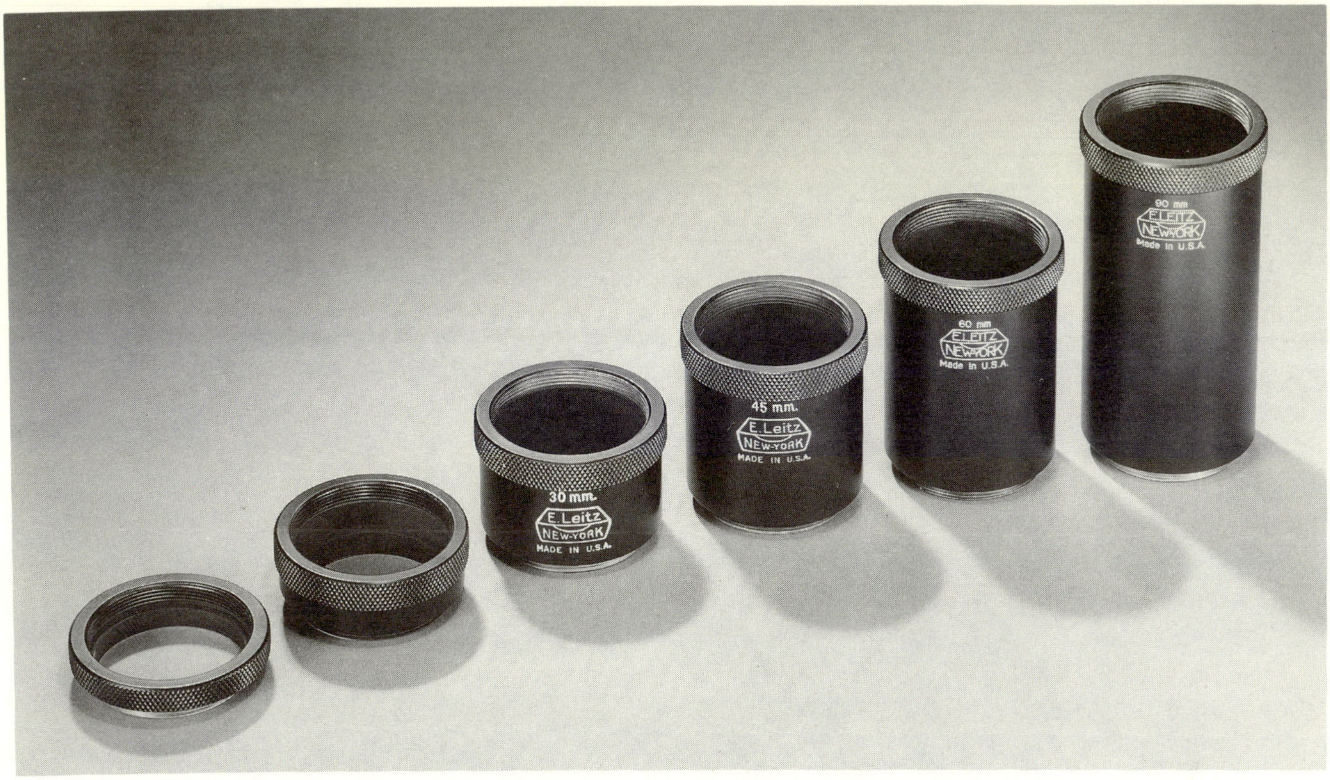

Extension tubes for Leica copying devices made by Leitz New York

Because of this demand, Leitz New York received a contract from the U.S. Signal Corps a few years later, to deliver 3,000 Leica outfits of the model IIIA with 3 each interchangeable lenses: 50, 90, 127 mm, plus universal viewfinder. The optical elements for these lenses came from Wollensack, Rochester, N.Y. Focusing mounts and rangefinder coupling-members were made in the shops of Leitz New York. Harman Kardon New York received the contract for the camera itself, but manufacturing difficulties delayed the completion of the contract , causing the Signal Corps to cancel the contract after the war with Japan ended.

Since the Signal Corps paid for the direct labor costs and the raw material, semi-finished parts now became Signal Corps property. These were bought back by Kardon with the intent of finishing the final product, utilizing a Kodak 50 mm lens. The camera was now called a "Kardon" and offered to the photographic dealerships.

Leitz New York was now changing its location again (1945). This time to 304 Hudson Street, a warehouse district, across the street from Heide's Chocolate Factory. The rent was much lower and the space more suitable for manufacturing and storage. The Kardon Camera was not readily acceptable to consumers, and Leitz New York had no interest in promoting it. By 1946, minimal importation of the Leica IIIa again resumed from Leitz Wetzlar.

During May 1946, John Slacks, (representing the A.P.C. Alien Property Custodian) and Alfred Boch flew to Berlin to consult with General Lucius Clay to establish a framework for the first post war importation of Wetzlar products.

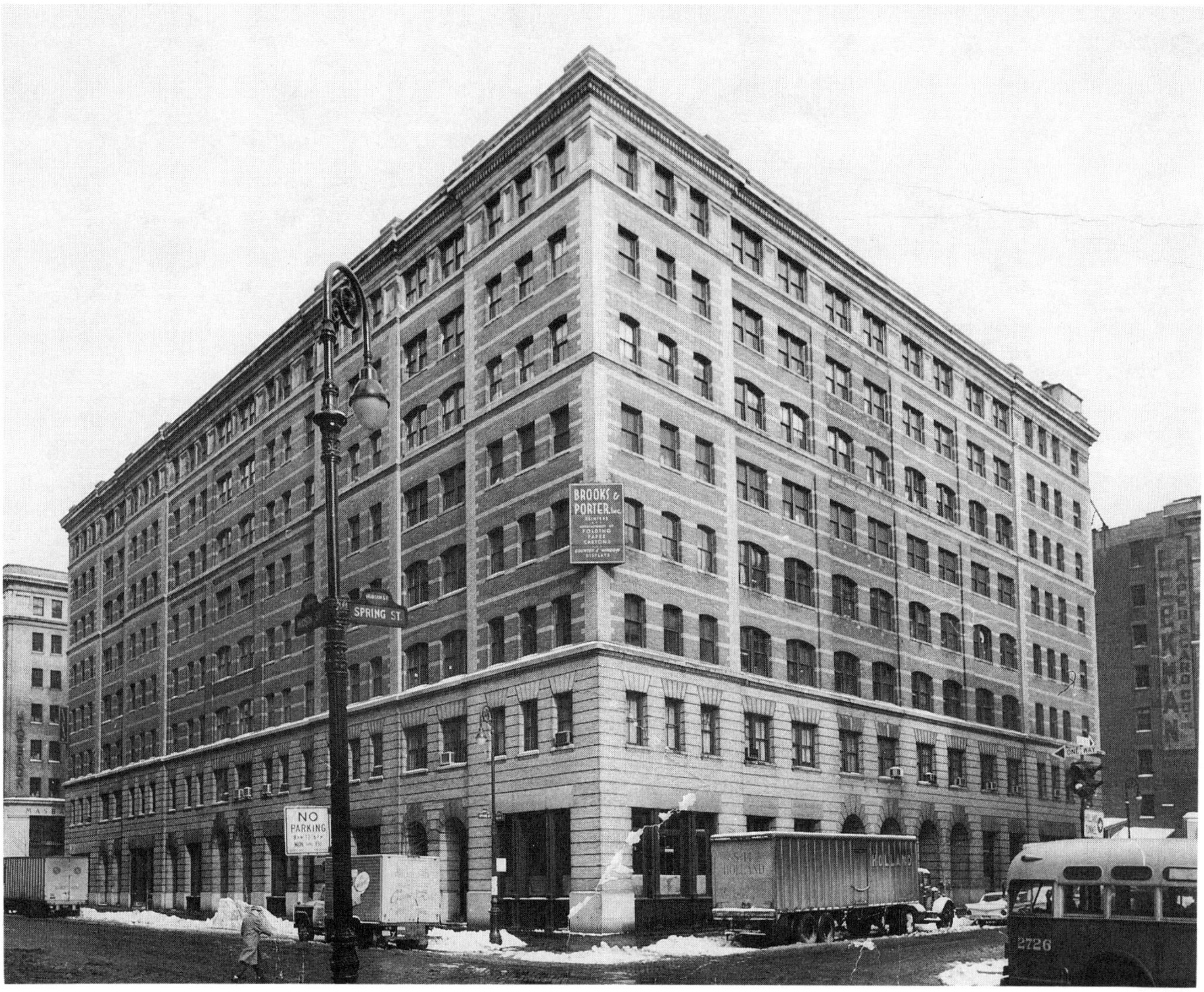

New location in 1945 at 304 Hudson Street, New York City

Chapter 6

LEITZ FROM 1946 - 1952

Leitz from 1946 - 1952

From 1946 to 1952 Leitz remained under the control of the Alien Property Custodian and survived by marketing instruments and camera products of its own manufacture. Leica cameras were imported to the U.S.A. in substantial numbers by returning service personnel and through the black market from German sources. American dealers quickly established direct relationships between themselves and their German counterparts , avoiding E. Leitz, Inc., New York. A letter from Balcom Trading Company, Inc., Tokyo, dated June 14, 1951, encourages American camera dealers to purchase Nikkor lenses for Leica cameras. Quote: "After months of waiting, we now have genuine Nikkor lenses of superlative performance proved by Life and other news photographers", says the above mentioned letter. Of course, the prices were far lower than equivalent Leica lenses.

Prices for new Nikkor lenses for Leica:

50mm f/1.4	Coated Lens	$198.00
50mm f/2	Coated Lens	107.00
50mm f/3.5	Coated Lens	54.50
35mm f/3.5	W.A. Ctd. Lens	89.50
85mm f/2 Ctd.	Lens w.Case	175.50

U.S. camera dealers, more than anxious to fill the pent-up demand, purchased used Leica cameras for resale causing Leitz's repair business to increase. That is not to say that many returning servicemen didn't retain their Leica outfits. Many enjoyed their new photographic hobby and became customers for accessories.

In April 1945, as the result of negotiations between Ernst Leitz III and Director Henri Dumur with the American Occupation Forces, approximately 30 Leitz employees were allowed to enter the plant with special passes. Their job consisted of cleaning up shattered glass and war debris.

American G.I.s then used their trucks to return to the plant specialized machinery and key devices for Leica production, from a bomb safe bunker where it was stored in 1943. Thus, the assembly of Leicas started, using left-over materials . By the end of 1945, Leitz employed 2,066 workers. The reason for the quick recovery of Leitz Wetzlar after World War II was in part because the major buyer of post war Leica cameras and lenses was the U.S. Occupation Army in Germany.

Meanwhile, in Leitz New York, most of the key people left the company. George Hesse, chief writer and editor of Leica Photography passed away at an early age. Gus Wolfman joined Photo Dealer as editor. Ken Poli established a career at Popular Photography. This author began to correspond with Tom Elwell, an editor of an English 35 mm journal, intending to bring him and his family to the U.S. to continue the most successful promotional medium Leitz had, Leica

Photography. This was achieved in 1947 and Tom went to work as planned.

In 1948, Mr. Slacks resigned from Leitz N.Y., to be replaced by Harold Horowitz, a New York City attorney, who became President and Chairman of the Board. Alfred Boch, likewise, had to resign. Wolfgang Zieler left earlier to start his own instrument business in the Boston area.

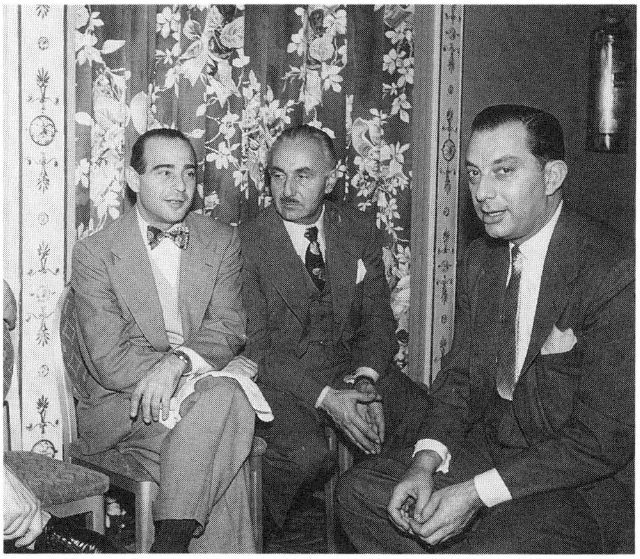

Harold Horowitz, President Leitz New York visits photo dealer in California (1950)

In October 1948, Managing Director, Dr. Henri Dumur, visited the U.S.A. to attend the Photographic Convention in Cleveland, Ohio. This gave him in opportunity to explore the status of his old agency, E. Leitz, Inc. New York, still under the control of the Alien Property Custodian. Dr. Dumur was contemplating the future of an enterprise which had faced numerous difficulties. Not only the state of affairs as an alien property, but the fact that Leitz New York was under new management (Harold Horowitz) once more, and the loss of trusted personnel (Alfred Boch). Furthermore, Leitz New York was still producing Leica accessories in competition with the factory in Wetzlar.

This author's own assurances expressed to Dr. Dumur in August 1948 seem to have persuaded him to continue the relationship. During a visit by Mr. Horowitz to Wetzlar, in the company of this author, toward the end of 1948, kept the agency agreement in tact with the provison that Leitz Wetzlar was not obligated to continue it after the company was sold by the U.S. Government.

Horowitz was a relatively young man, about 45 years old, ambitious, with connections to the Department of Justice, which managed the Alien Property Custodian. Leitz was foreign territory to him and to run a business specializing in scientific instruments and miniature cameras, he needed help from old employees. Leitz New York was one of the few properties not yet up for sale by the Alien Property Custodian. In fact, it remained at the bottom of the list until 1952. The only new key employee during that period was Stanley Samuel, an acquaintance of the new President. The service personnel, camera and microscope repair technicians, ,who remained with the company, in spite of the changes, were engaged in

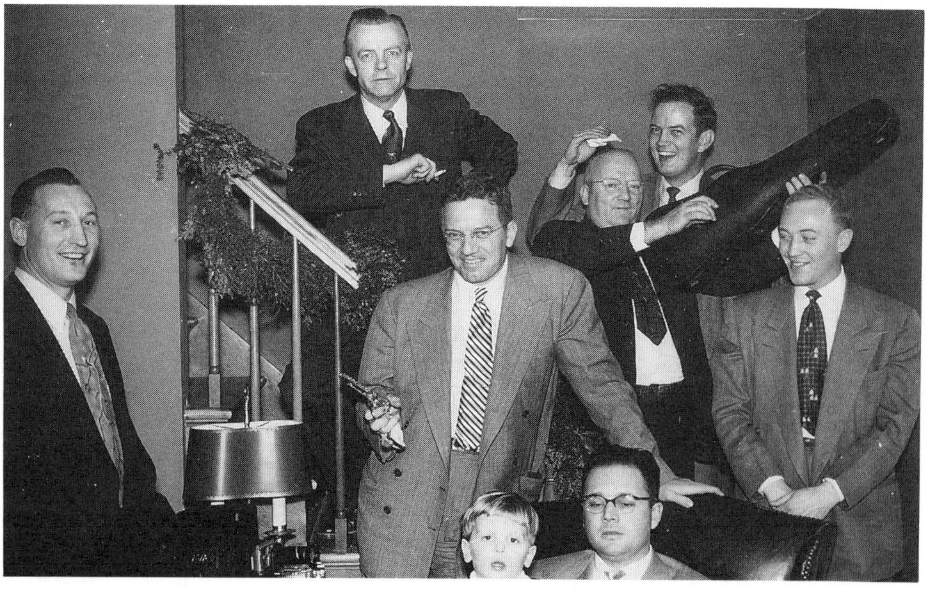

Group of Leitz/Leica Salesmen in 1952

the manufacture of accessories, colorimeters, carbon meters, and other products.

By 1948, the German Mark was re- established and the economies of the various occupation zones began to recover, Horowitz decided to travel to the Leitz factory in order to obtain delivery commitments. He was ready to place orders which could be delivered in monthly quantities over a period of 12 months. The reception in Wetzlar was cordial, taking place in the conference room of the old office building near the Kalsmunt gate.

To describe the meeting as cordial misses the underlying tone of Dr. Henri Dumur, the spokesman for the organization. Dumur was a scholarly, soft spoken business man and negotiator. His English was accented with his native Swiss French. Horowitz and this author were not given a tour of the factory, but certain delivery promises were made and kept, so that by 1949 a limited flow of Wetzlar products resumed. By 1951, the quantities of Leica Cameras sold were as follows:

1951 Sales :

To Germany (Post Exch.,etc.)	4,489	=	32.48%
To Export without the U.S.A.	23,247	=	52.12%
To Export U.S.A.	6,807	=	15.40%

Accessory lenses :

Germany	24,172	=	34.30%
Export without U.S.A.	36,716	=	52.09%
Export to U.S.A.	9,595	=	13.61%

The 1952 U.S. retail price list again shows Leica models If, IIf and IIIf. The IIIf with Elmar 50 mm F 3.5 lists for U.S.$ 278.50. The assortment of available interchangeable lenses included the Hektor 28 mm, the Summaron 35 mm F 3.5, Summitar, Summarit 50, Summarex 85 F 1.5, Elmar 90, Hektor 135, Mirror Reflex Housing and others. Viewfinders for all types of lenses were available, as were filters, lens caps and carrying cases. Darkroom equipment followed, together with Leica meters, near focusing, copying and reproduction devices. Valoy and Focomat enlargers as well as prism binoculars completed the offering in limited quantities. The Leica Manual was again available and "Leica Photography" resumed publication.

During 1951, the total number of microscope imports to the U.S. with a value of $ 25.00 was:

County of Origin	Qty	Value $
Germany	5,237	472,335.00
Japan	1,563	101,030.00
United Kingdom	522	39,309.00

In the following year the quantities became substantially higher. Rather than hire new personnel without technical training, Leitz New York now began to establish regional dealerships for scientific instruments. A basic inventory was required and all inquiries received by the company were now forwarded to the local distributor. A stocking dealer was required to invest $ 8,000, represented by the following instruments: Leitz Panphot for biological laboratories, binocular magnifiers for college classrooms, micro-projectors for hospitals, metallurgical microscopes, polarizing microscopes for geology departments and the U.S. Geological Survey, camera attachments, and pathologists microscopes.

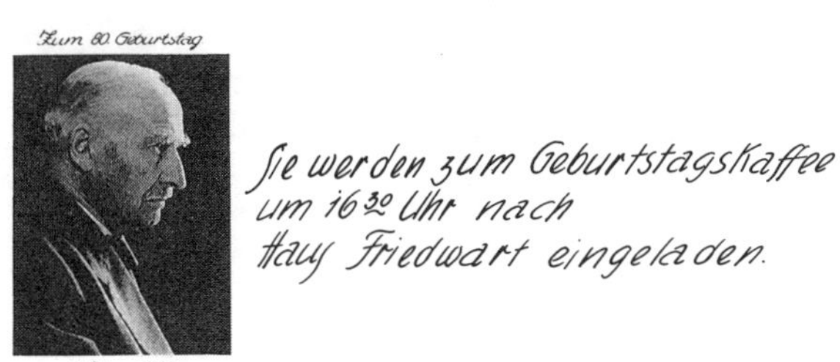

Invitation to attend the 80th birthday of Dr. Ernst Leitz II at Haus Friedwart on March 1, 1951

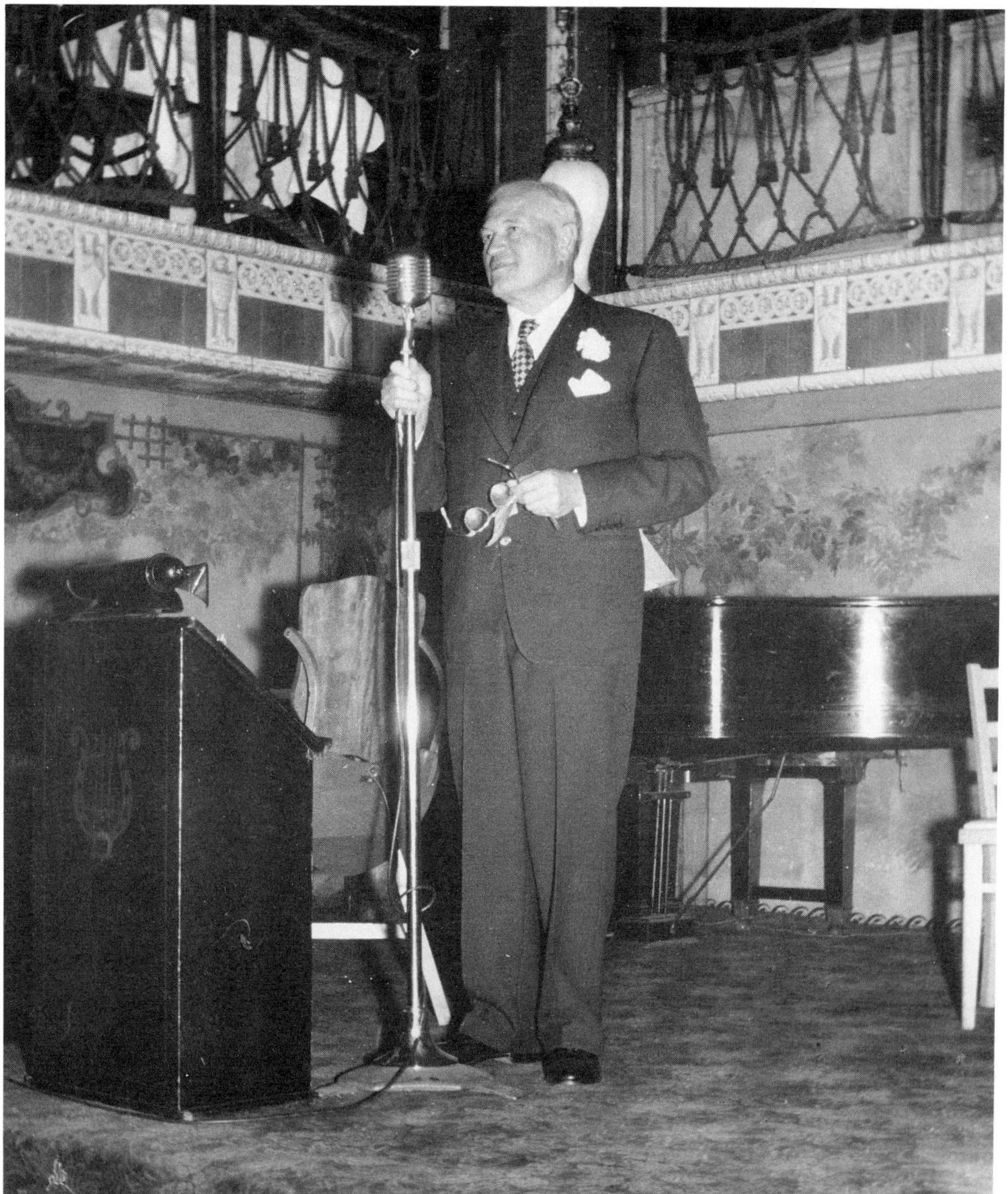

Henry Mann

Chapter 7

THE SALE OF LEITZ NEW YORK

The Sale of Leitz New York U.S.A.

On Friday, June 13, 1952, E. Leitz, Inc., New York, which had been taken over by the Alien Property Custodian from its German owners in 1942, was offered for sale to the highest bidder. The original owner, seven years after the end of World War II, was not permitted to bid. The new owner had to be a U.S. citizen and the proceeds were used to reimburse Americans whose property had been confiscated during the war.

400 capital shares were offered and the highest bid was rendered by Melvin T. Berry of Providence, Rhode Island. His bid was $ 677,779.00. In 1951-52 Leitz New York had realized a net profit of $ 107,300.00 on sales of $ 3,829,070.00. The Alien Property Custodian rejected that bid as too low and readvertised the sale at the Alien Property Office in New York, 346 Broadway, on July 23, 1952.

The high bidder this time was Dunhill Corporation, 660 Fifth Avenue, New York City, Mr. Arthur Mainzer, President, assisted by Mr. Dunston. The amount of the bid was $ 787,008.12. other offers and their bid amounts follow:

Charles Besseler, 60 Badger Street, Newark, N.J.
$ 782,000.00

Manca, Inc., 63 Wall Street, New York City
$ 721,770.00

Robert R. Hammond, 281 Madison Ave., New York City
818,799.00
Disqualified because one year payment terms were requested.

Fulton Gold Refining Corp., New York City
$ 250,000.00

Samuel Ellison, 229 W. 36th Street, New York City
$ 1,000.00
Disqualified since no down payment was offered.

The award was made August 6th . On the following day Mr. Mainzer came into the Leitz office at 304 Hudson Street, New York City, to take over as president. The assumption was that Dunhill wanted to sell Leica cameras through their retail stores.

Communications with the Leitz factory in Wetzlar followed and to Mr. Mainzer's surprise, the German management, represented by Dr. Henri Dumur, did not agree with the sale of their U.S.A. distribution center to Dunhill. They stated flatly that while they had no control over the sale of their property, they were not obliged to continue the agency relationship with the new owners. Mainzer sent a telegram to Wetzlar which mentioned "how happy he was to have succeeded in purchasing E. Leitz, Inc., New York, and that he was looking forward to a long and fruitful relationship with the factory in Wetzlar". Wetzlar's response must have seemed unbelievable to Mr. Mainzer.

As one who was in Leitz New York management during 1952, this author recollects the enormous difficulties to keep the business going. The company was literally breaking apart. Employees resigned, especially those in the production and service areas, as it appeared that

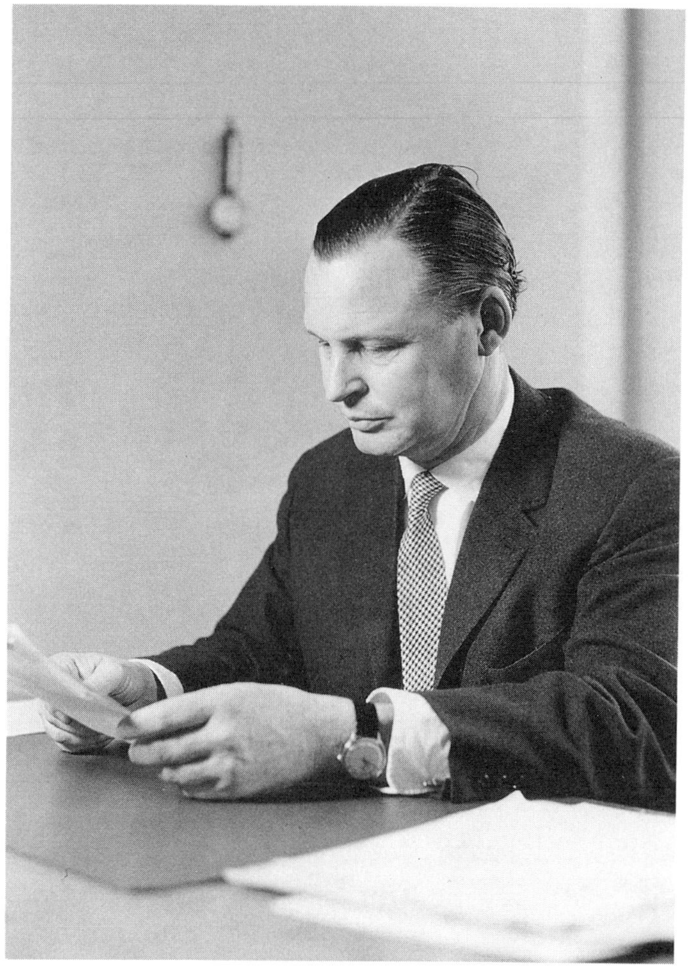

Guenther Leitz

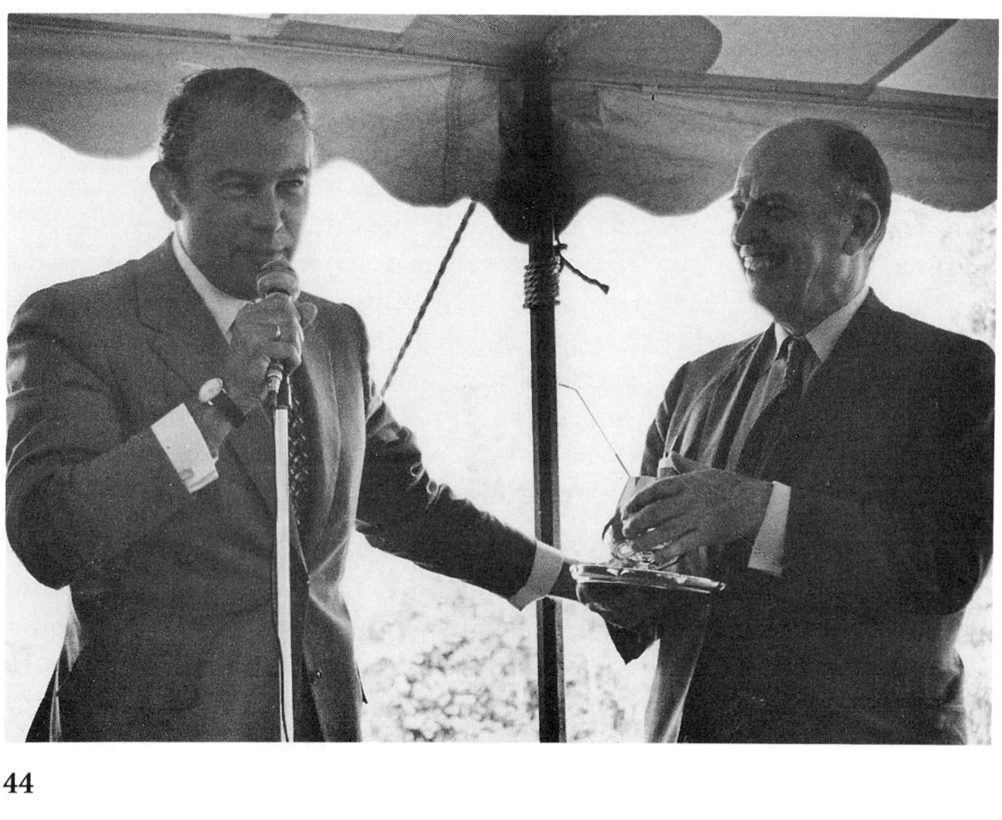

James Scarff and William Mann

there was no longer a future at Leitz New York. Even Dunhill had to accept the fact that the current inventories and production facilities would soon be exhausted, and that the German factory had sufficiently recovered from the war so that Leitz Germany could appoint anyone to represent them in their most important foreign market, the U.S.A..

However, having mentioned all the negatives, Dunhill did purchase the trademark 'Leica', which was registered in the name of E. Leitz, Inc., New York, at U.S. Customs. A valuable asset which, under certain circumstances, could force the Leitz factory to change the name of their most important product, the Leica, at least as far as the U.S. market was concerned.

Mr. Mainzer, as new trademark owner, immediately took the appropriate steps to preclude the use of the name 'Leica' by filing a complaint with the Bureau of Customs on September 10-12, 1952 with respect to publications such as Leica Fotografie, Frankfurt, Germany and Leica Photography, London, England.

On August 9, 1952, Leitz Wetzlar informed Mr. Mainzer that they would ship pending orders against sight draft via Pan American Airways or Seaboard & Western.

On August 29, 1952, Leitz Wetzlar informed U.S. customers by circular letter: "You can be without any fear for the further distribution of our products in the U.S.A.. We do not intend to give up any part of our American business which has been so successful for the last 60 years. Our old firm of E. Leitz, Inc., having become the prey of outsiders who did not understand that we would not be ignored, left only one possibility to us: the erection of a new and independent agency. We beg you to be patient for a little while. Our new agency will soon be able to give you the very best service!"

Manca fortunately had assured themselves of a representative agreement with the factory, an agreement which covered 10 years, subject to renewal after that time frame expired. After negotiations among the two bidders, Mr. Arthur Mainzer and Mr. Henry Mann, Dunhill decided to sell to Manca, at a profit to be sure, but by the end of 1952, E. Leitz, Inc. was in the hands of Henry Mann, this time with the blessing of the Wetzlar factory.

The immediate question now coming to mind is: What is Manca? Manca was and is a Securities Investment firm at 63 Wall Street, New York, owned by Henry Mann, Gracie Square in New York City. Mr. Mann was known for his pre-war dealings and financial expertise with Germany. He contacted the factory through intermediaries to determine whether he could count on an exclusive agency agreement if his bid was awarded to him. It was just $ 66,000 below that of Dunhill. He chose the

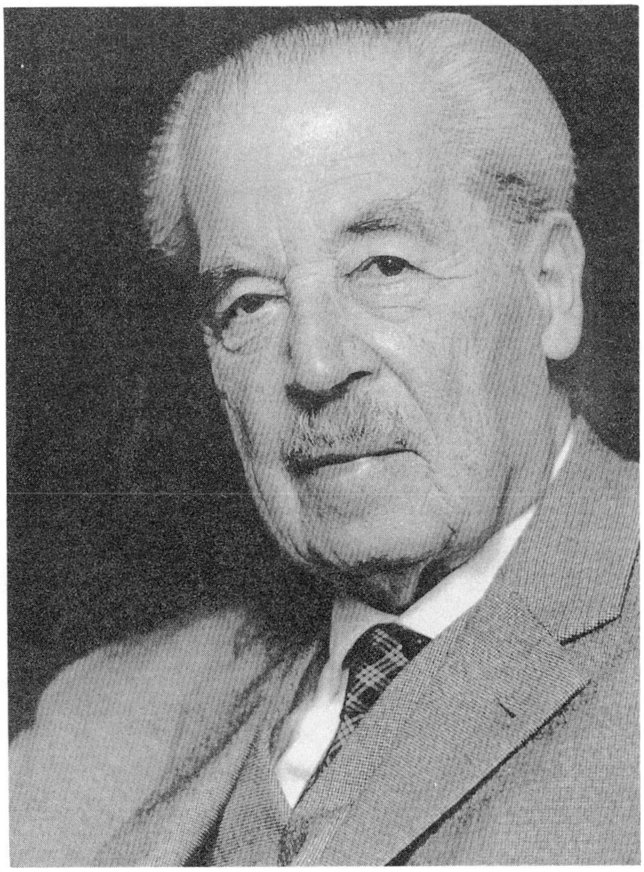

Dr.Henri Dumur

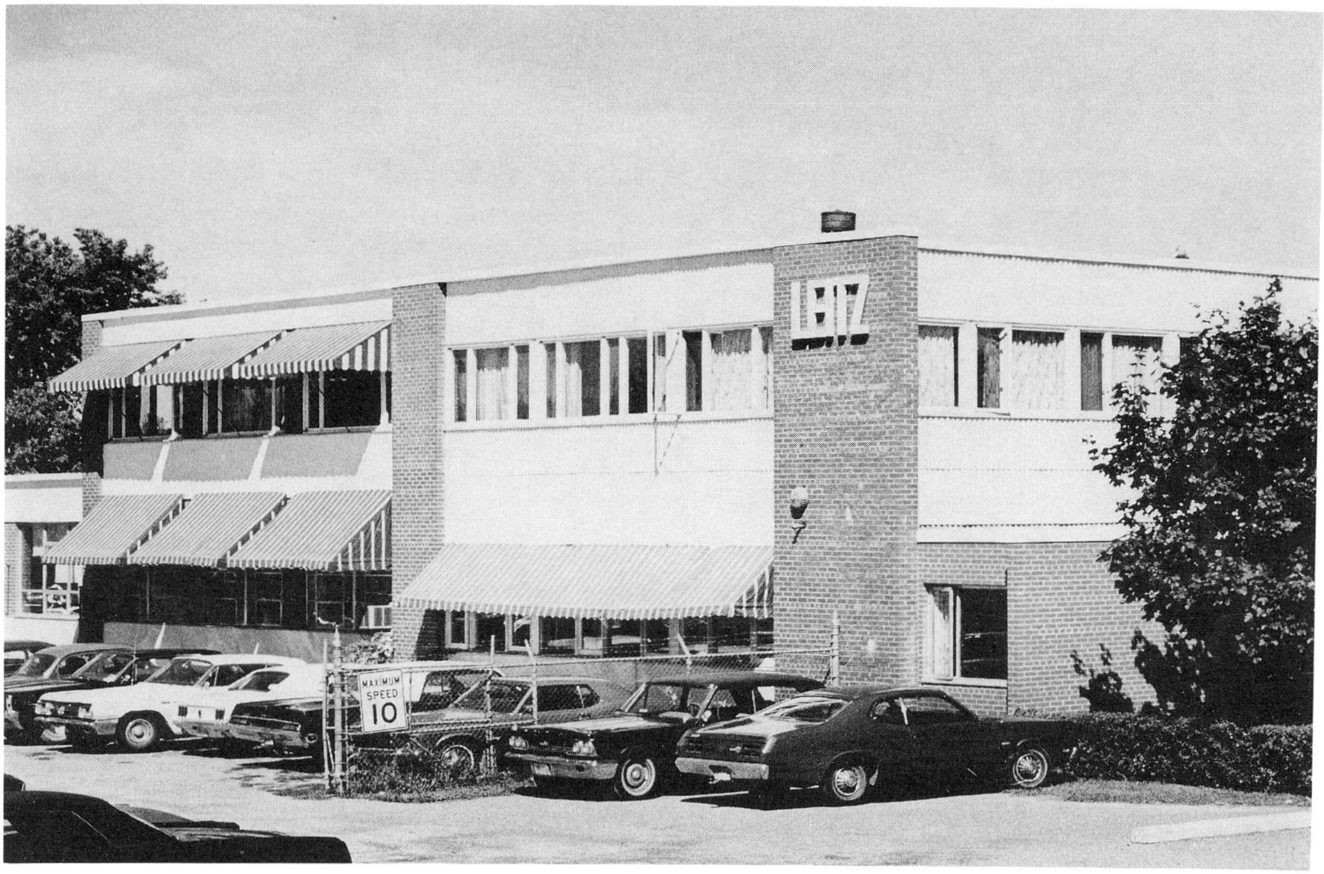

Leitz plant Midland, Ontario

name Manca for the same reason that Leitz chose 'Leica'. Henry Mann was born in 1890, came to the U.S.A. in 1913 and became a U.S. citizen in 1920. Between 1917 and 1926, he was controller of the Crown Cork & Seal Company in Baltimore. As Director of the American Chamber of Commerce in Berlin in 1926, he established many connections with German industry and banking. He returned to New York after the outbreak of World War II and established his own securities corporation.

While Mr. Mann's background was in itself sufficient to provide the confidence Leitz wanted for its future business connection in the U.S.A., the factory also requested that its own technical experts be re-hired and made part of the new agency, a stipulation Mann readily agreed to.

When the new Leitz Agency, E. Leitz, Inc., 468 Park Avenue South, was organized, Mr. Mann not only brought Mr. James Scarff and George McCaffrey into the organization, he also re-hired Alfred Boch of the pre-war Leitz U.S.A. organization, an executive who had been with that company since 1935. He also hired this author, who since 1936 specialized in sales of scientific instrumentation. Mr. Mann's son, William, likewise took on a marketing position with his father's company. From January 1956 to July 1957, he went to the Leitz factory in Wetzlar to become familiar with their extremely diverse manufacturing program, broadly covering science and industry, as well as 35 mm photography. Messrs. Scarff and McCaffrey were experts in finance and accounting respectively.

Manca and Leitz Wetzlar were aware of the problem of the trademark registration at U.S. Customs. Quite a few cameras, brought into the U.S.A. at that time by private citizens had to have the name 'Leica' obliterated. In fact, Wetzlar threatened to change the name altogether, a threat which undoubtedly contributed to the sale of E. Leitz, Inc. from Dunhill to Manca.

In March 1952, the Free Press of Midland, Ontario, Canada, announced that "the selection of an Ontario Industrial Site at Midland, where a branch factory will be erected by the world famous Ernst Leitz Company was announced by the Hon. William Griesinger, Ontario Minister of Planning and Development".

Ernst Leitz III conducted location studies in Ontario, Quebec and the Maritimes before selecting the Georgian Bay town of Midland. Midland's curling rink became the first home of Ernst Leitz (Canada) Ltd. Walter Kluck arrived in April 1952, followed by Guenther Leitz, the youngest of the three Leitz brothers, in May, together with the first group of expert technicians and their families.

Guenther Leitz was most anxious to establish a foothold on the American continent. While narrowly escaping the Russian advance in Rumania during 1944-45, he also took seriously Stalin's threat to take over all of Germany . The growth and development of the Leitz Canadian plant deserves a full expose, as does Guenther Leitz's early death in 1969. However, this does not fit into the story of Leitz U.S.A.. The Canadian venture is mentioned here because it took place at the same time when Leitz U.S.A. was up for sale.

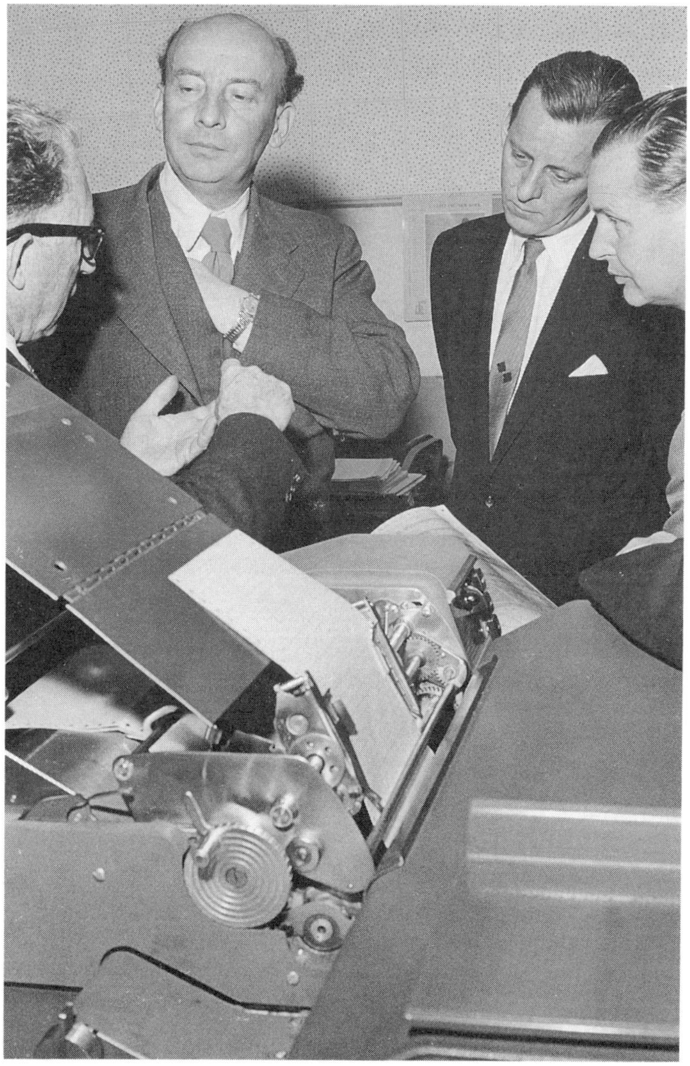

George McCaffrey demonstrates parts of the 360 IBM computer to Dr. Ernst Leitz III and Guenther Leitz

Chapter 8

YEARS OF PROGRESS

Years of Progress

First, however, a brief comment on how the Leitz Works in Wetzlar fared when General Hodges of the U.S. First Army occupied the town and its plants on April 1, 1945. Since Wetzlar was not being defended, the Leitz plant was turned over to the military without resistance. In fact, Dr. Ernst Leitz II stood at its gate and let the American soldiers in.

Subjected to a minimum of air raids during the war, the plant suffered minimal damage. The important machinery had been stored underground and a limited cadre of skilled workers was still operating. Glass and coal to heat the workshops were the major necessities for continuing operations. The occupation forces lost little time to help the Leitz Works get operational producing microscopes, repairing army binoculars, and completing Leica cameras. Glass and coal were available in the British sector of occupied Germany and barter became the method of exchange. Leica Cameras were exchanged for deliveries of coal and glass to replace broken windows.

Toward late 1945, a limited number of Leica cameras were again in production. In 1946, the exportation of cameras resumed.The major purchaser was the U.S. Post Exchange. By 1948, after the new German Mark was introduced, Dr. Dumur, then one of the managing directors of the Leitz plant, persuaded the Military authorities to allow limited exports to permit Leitz to maintain its marketing position in the world market.

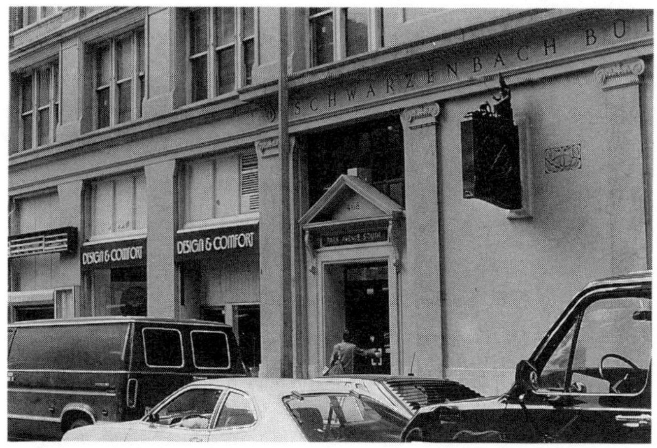

468 Park Avenue South, New York City

With Henry Mann taking over the leadership of E. Leitz, Inc. New York, in 1952, a major expansion took place. While 304 Hudson Street retained the warehousing and repair functions, the office moved to 468 Park Avenue South. A new accounting and inventory system was introduced, a computer purchased, a new advertising agency hired, and the contacts with the factory fortified. The demand for Leitz products continued to be strong, in spite of the war, with pent-up demand in all sectors of the economy. The reputation of the Leica camera had actually improved.

The new management made a great effort to reach out to its customers and dealerships as quickly as possible. The difficulties of the previous several years had been overcome. The public seem to accept the new management team. Meanwhile, both Leitz management, and the

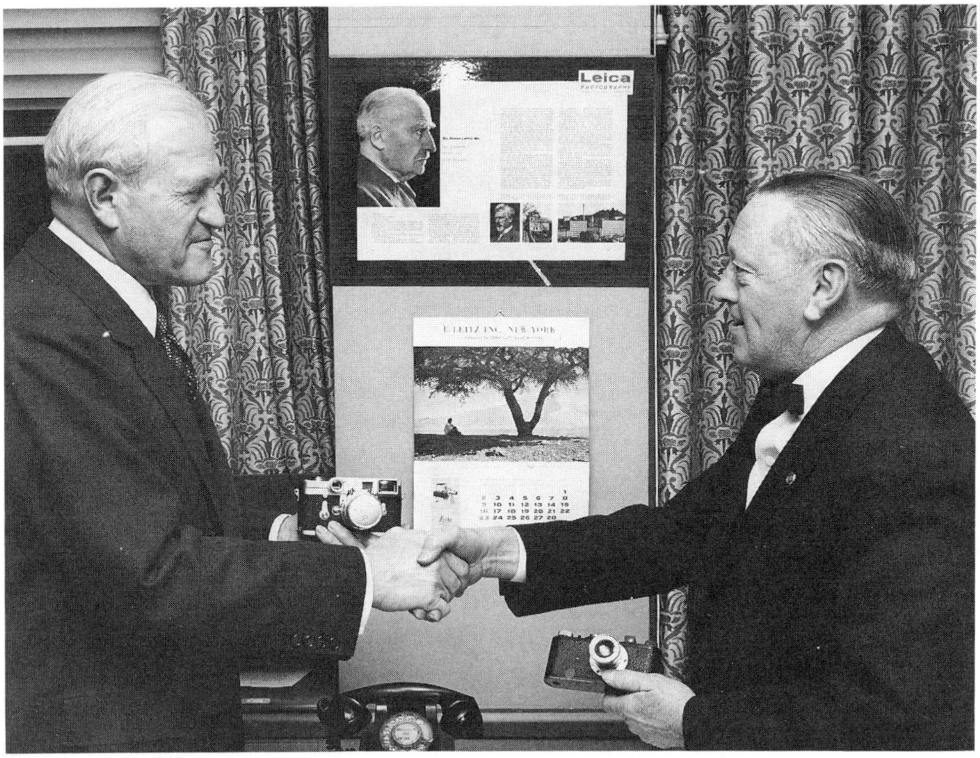

Henry Mann presents Leica camera with significant serial number to Bud Wichert of the National Geographic Society, Washington, D. C.

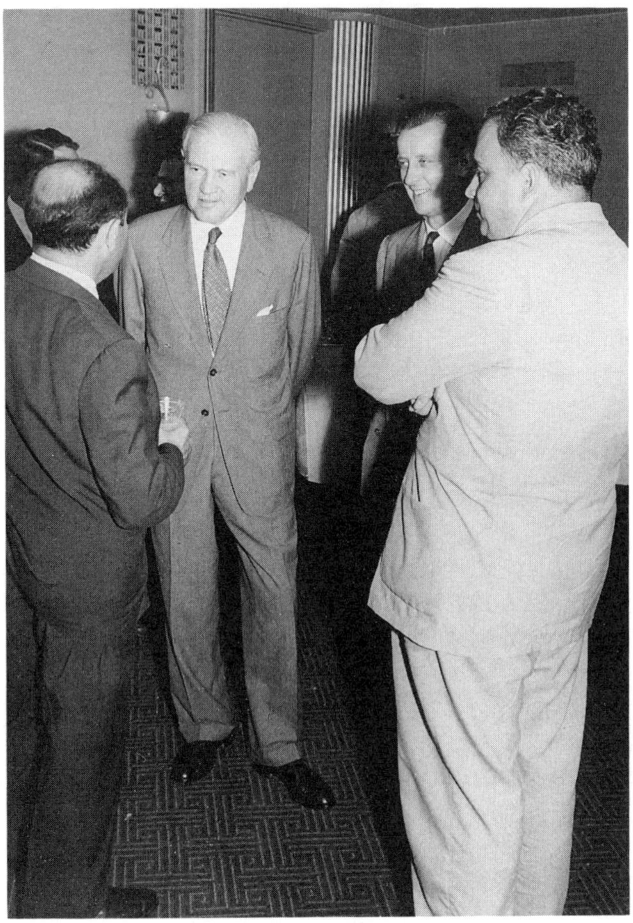

Henry Mann speaks with Alfred Eisenstaedt during a reception

factory in Germany worked hard to establish a solid basis for the distribution and service of its world renowned products, Leitz microscopes and Leica cameras.

The customer was again king. Both, a Leica School and a Laboratory for Applied Microscopy were reinstituted immediately. Leica dealer salesmen were trained to explain the Leica system of photography. Instrument dealer personnel sharpened their skills in dealing with the intricacies of the broad base and diverse scientific instrument program. John Brooks was rehired as head of the Leica School and Dr. Peter Bartels, after extensive studies at the factory, became the expert lecturer for Leitz scientific instruments. Sales meetings for scientific instrument and camera representatives became annual events during January of each year. The closing dinner party for each meeting was held at the Knickerbocker Club on 62nd Street in New York. A gala occasion indeed.

Branch offices had to be re-established, equipped with samples of the latest instruments. Regional representatives were trained and sent into the field or assigned to Leitz offices in Washington, D.C.; Chicago, Illinois; Houston, Texas; San Francisco or Los Angeles, California.

The Mann family retained "Manca" as the importing arm for Leitz. During the excise tax period, it was beneficial to pay tax on the imported value instead of on the net to dealer value. However, even after the excise tax was rescinded, Manca became the "import expert". It had developed a group of skilled experts, dealing with the complications and technicalities of customs laws.

Meanwhile, Leitz Canada started operations. Guenther Leitz visited New York and explained during a sales meeting what to expect from the new Leitz plant in North American.

Other changes were taking place in the market. Competition from Japan was emerging. Julius Huisgen tested the Canon lenses and found them very good, especially at wide apertures. Similarly, Japanese microscopes entered the U.S. market. Their prices were astonishingly low. In medical schools, students no longer purchased their own microscopes. The schools purchased the instruments and rented them to their students. Leitz New York succeeded in obtaining many such contracts. Yet, the market had permanently changed, and Japanese competition in the field of optics had to be reckoned with.

Sales figures increased as the new initiatives began to pay off. Prosperity was not limited to the Leica business which prospered. By the mid 1950's the electronics industry started to assemble circuit boards for radio and television units, which required large numbers of optical inspection and quality control instruments. Practically all manufacturers, domestic, Japanese, and European, benefitted from the demand. While the market grew exponentially, the competition for market share grew at the same rate.

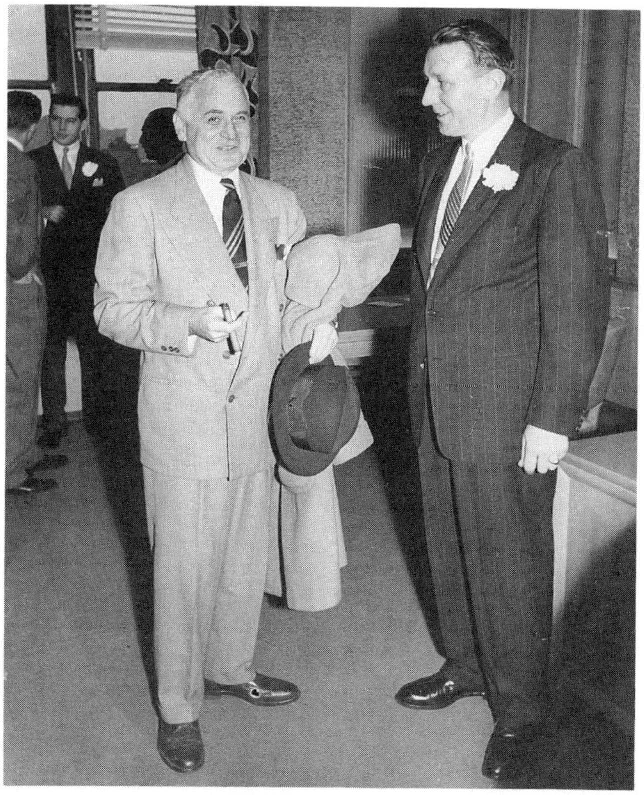

Alfred Boch with longtime New York camera dealer

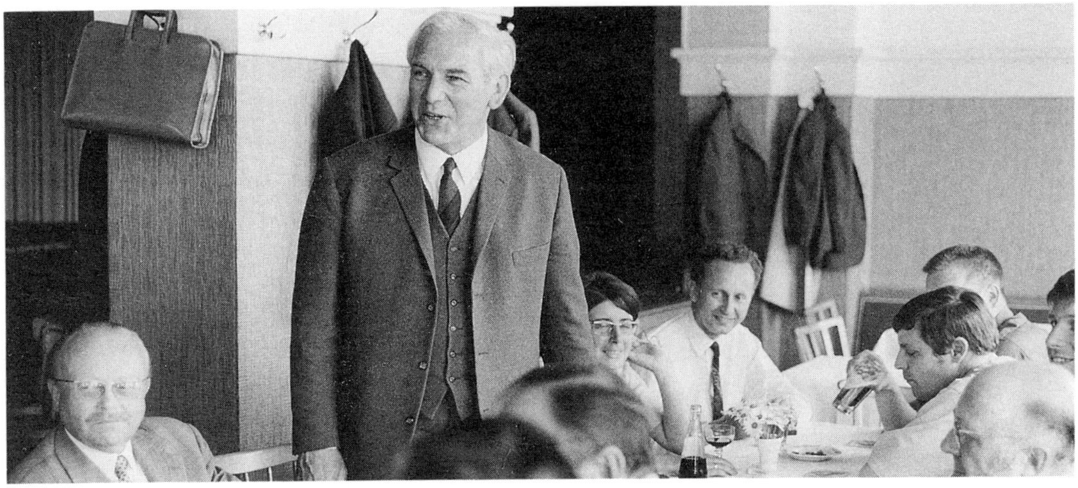

Leitz NY representatives were given a tour of the Leitz plant for their achievements and dined German style. Director Glier addresses the group

William Mann and Karl Jung talk to a potential customer, Prof. Dr. George L. Wied, University of Chicago, at the new Leitz Branch Office in Evanston, Illinois

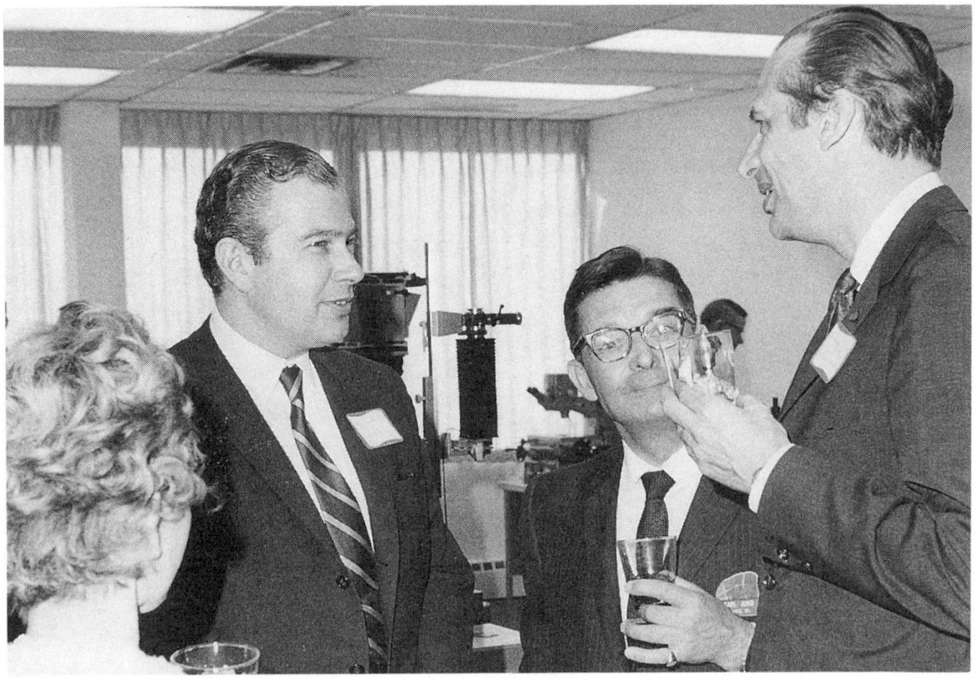

The strength of Leitz lay in the new design of compound microscopes and measuring stages for the inspection and quality control of wafers, grown from silicon crystals. Computer chips were now cut from three inch diameter silicon wafers. The rejection rate was more than 30%. Firms such as I.B.M., Motorola, Texas Instruments, to name just a few, equipped their laboratories with large numbers of such instruments in order to improve output. These instruments are still in use today even though the diameter of a wafer has since grown to 8" and output has increased.

A listing of a selected group of Leica equipment and Leitz microscopes sold by E. Leitz, Inc., New York, and imported from Wetzlar between 1950 and 1953 is shown in the following table.

	1950	1951	1952	1953
Medical Microscopes	1,225	1,452	1,126	822
Research Microscopes	84	109	95	97
Industrial & Specialty Microscopes	87	103	177	120
Leica Cameras 3f-2f-lf-3c-2c-lc (**) with 50 mm lenses	3,553	6,068	5,726	7,501
Leica lenses, incl. Wollensack 90 and 127 mm (*)	2,795	2,861	3,904	5,765

(**)After 1952, Leica cameras lc and 2c no longer available.
(*)After 1953, Wollensack lenses were no longer supplied.

In the early 1950's, Japanese competition was gaining market share. Both Canon and Nikon offered cameras and lenses at lower prices and, more importantly, with improved correction characteristics. Leitz replaced the Summitar with the Summicron 50 mm F/2 which more than made up for any deficiencies that might have been noticeable. Its design was based upon characteristics of new glass invented by Weissenberg at the Leitz Wetzlar glass laboratory. Both dispersion and refraction characteristics of the new glass assisted the lens designers to improve color rendition and detail sharpness.

None of these improvements however could compare to the Leica M3, shown at Photokina in Cologne that year for the first time. This camera type had been "in the works" since 1935. Slightly larger than the Barnack models, it had a bayonet mount for the rapid change of interchangeable lenses. The built- in rangefinder base was longer, the viewfinder automatically indicated the viewing frame for the lens on the camera, and many other first time features had been incorporated. The competitors, both domestic (primarily Zeiss) and Japanese were faced with a new challenge and Leica sales again increased tremendously all over the world.

Note: By April 1953, the new Summicron 50 mm F/2 was supplied instead of the Summitar for all IIIF cameras. Peerless Camera in New York City purchased 450 Leica lf and 2f bodies to be sold with foreign lenses in 1953.

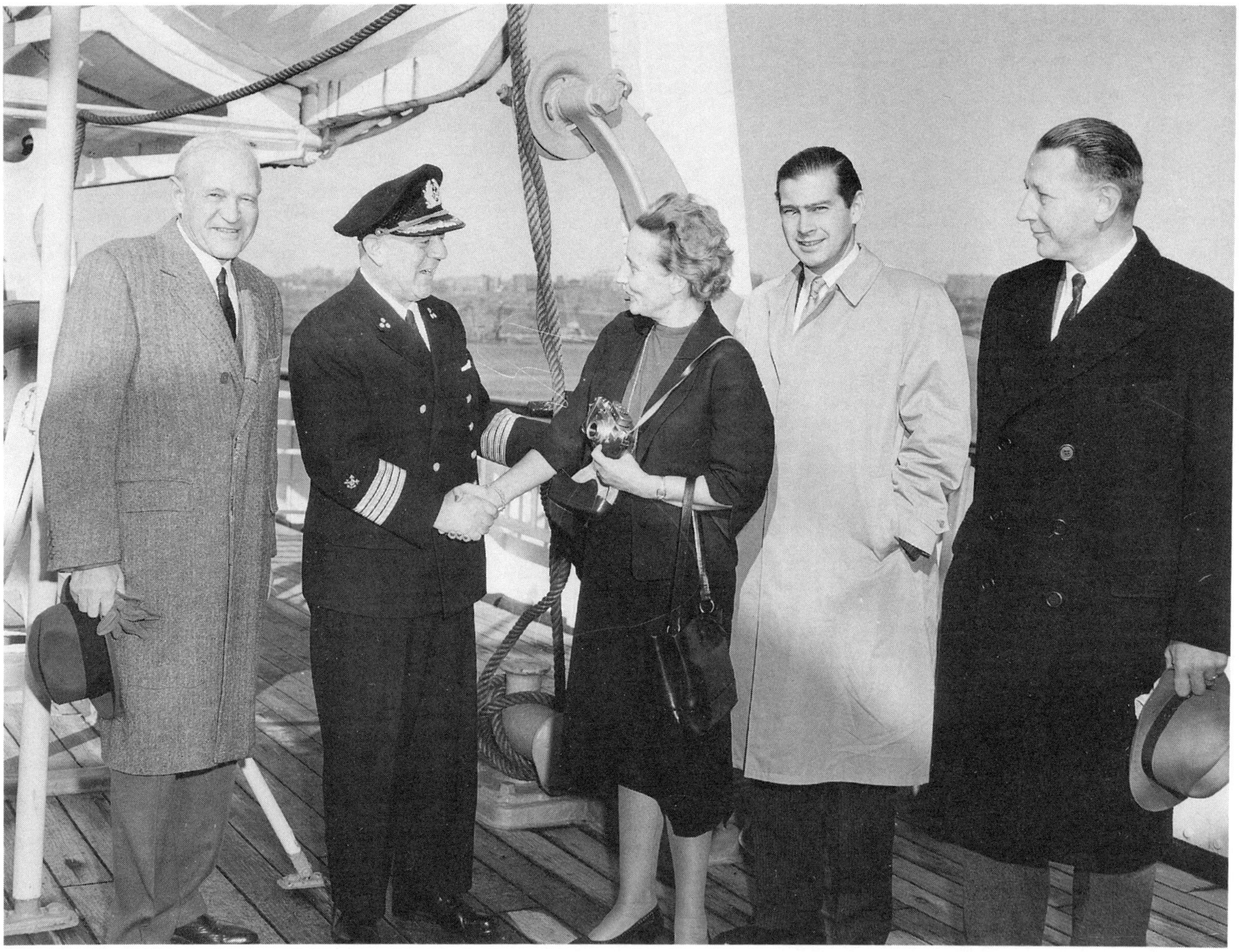

In 1957, Dr. Elsie Kuehn-Leitz returns to Germany after visiting twenty-eight states in U.S.A. From left to right: Henry Mann, Capt. Lorenz, Elsie Kuehn-Leitz, William Mann, Alfred Boch

Chapter 9

HENRY MANN'S FIRST 10 YEARS

The First 10 Years of Henry Mann's Ownership of E. Leitz, Inc., New York, 1952-1962

Any radical change, such as occurred when the Alien Property Custodian relinquished ownership and management of the firm to Henry Mann, requires time to settle before showing positive results. It did not take long at all, however, due to the many positive steps the new management took and because of the proliferation and availability of key products from the factory in Germany. This change can best be illustrated by the following table of the quantities of cameras and lenses sold:

	1954	1955	1956	1957
Leica M3	6,377	5,426	6,082	4,549
Leica 2f	1,143	1,567	1,393	170
Leica 3f	2,505	2,232	1,791	127
	10,286	9, 7@5	9,613	
Leica 3g				3,775
				8,914
Leica Lenses all types	16,351	16,354	16,137	17,115

In 1954, retail prices for Leica cameras were reduced. A few examples follow:

	Old	New
Leica IIf w/o lens	$ 131	120
Summicron 50/F2	210	180
Leica IIIf with Summicron 50 mm F2	393	348
Leica If with Elmar 50/F3.5	143	132

Quantities of cameras and lenses sold 1958 through 1962 were as follows:

	1958	1959	1960	1961	1962
Leica M3	2,943	3,431	2,864	1,961	1,806
Leica 3g	2,142	1,237	672	88	
Leica M2	2,115	2,465	1,872	1,500	1,289
Leica lg	391	379	409		
	7,591	7,767	5,823	3,717	3,475
Leica Lenses all types	13,877	16,397	8,753	6,704	5,502
Leicina 8S				1,192	467

The Leica M2, a slightly simplified version of the M3 camera,was introduced in 1958 for the first time, followed in 1960 by the Leicina 8S. The Leitz plant in Germany had expanded its administrative and manufacturing facilities and began to feel financial pressures.

Microscopes were promoted by advertisements in practically all of the biomedical journals, such as Science, and to specialized disciplines like geology, botany and parasitology. Industrial instruments were featured in Metals & Alloys, Industrial Engineering Chemistry and the new electronics journals. Print advertising aside, probably the most effective promotions were product exhibits during conventions by the scientific and industrial communities. E. Leitz, Inc., New York, also became a member of

S.A.M.A., the Scientific Apparatus Manufacturers' Association, which maintained an optical section represented by, among others, Bausch & Lomb, American Optical Company, Nikon U.S.A., Olympus U.S.A., etc. Annual meetings gave Leitz the opportunity to compare notes with its competitors and discuss dealer relationships.

Among the unique instruments of that time was the Leitz nuclear track microscope (1959-60), of which the Los Alomos, New Mexico, laboratories were the principal users. Observation was made at a magnification of 2500x and determined the scattering of nuclear tracks in photographic plates. A glass rule on the mechanical stage featured a scale indicator which not only showed if the specimen was properly in focus, but it also served to accurately measure and locate tracks of 0.01mm in both x and y directions.

It was also necessary to examine the condition of fuel rods in nuclear reactors. Leitz designed the Metallograph MM5RT which allowed the optical path to pass through lead walls, permitting the observation of these rods by remote control.

As far back as 1949, the E. Leitz, Inc. budget for promotional activities, publicity, etc. to cover both the scientific instruments as well as photographic market amounted to between $ 35,000 and $ 50,000 per month. By 1962, these promotional activities had more than tripled. The camera ads were placed in the most widely read magazines such as Life, Colliers, National Geographic and, of course, in the popular photography magazines.

Leitz's photo business began to lose market share, mostly due to Japanese competition, but also because labor costs in Germany had risen a great deal and skilled labor was hard to find.

The introduction and acceptance by the public of single lens reflex cameras by Japanese manufacturers in the early 1960's did not go unnoticed by the Leitz New York sales department. In fact, Japanese SLR's gained so much market share that many Leitz friends, and the photographic press, urged Leitz to consider building the Leicaflex and to compete with the Japanese in the growing SLR market.

Toward the end of 1957, Bob Schwalberg, a senior photographic reporter for Popular Photography expressed the desire to leave that journal, following visits to Photokina in Cologne. He spent some time in Wetzlar to exercise his influence on camera design as well as to improve instruction

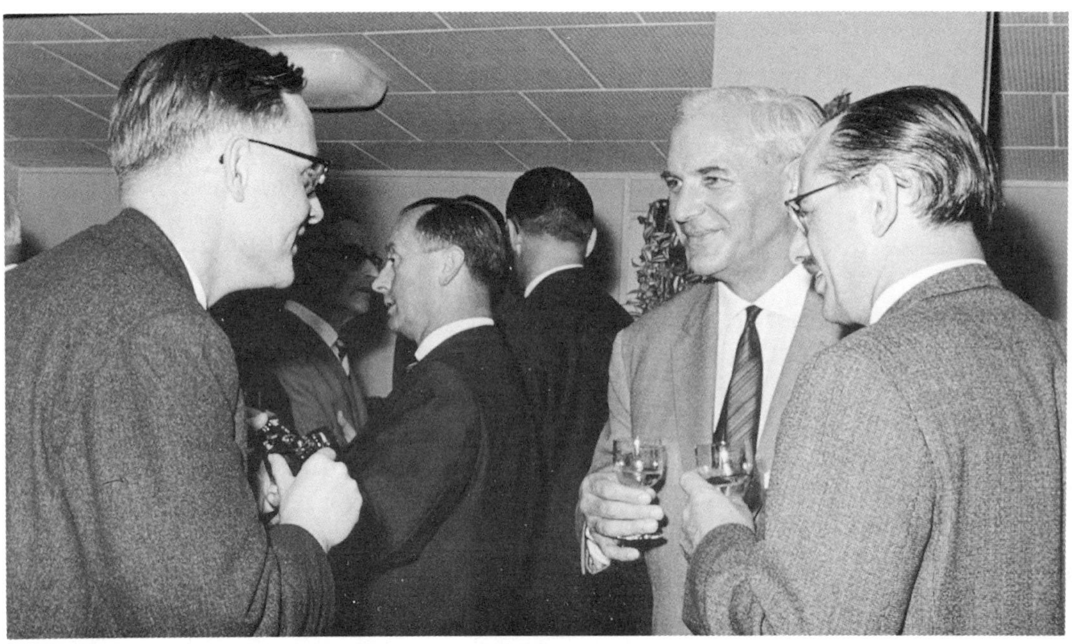

Bob Schwalberg proposes ideas to Director Glier at Wetzler

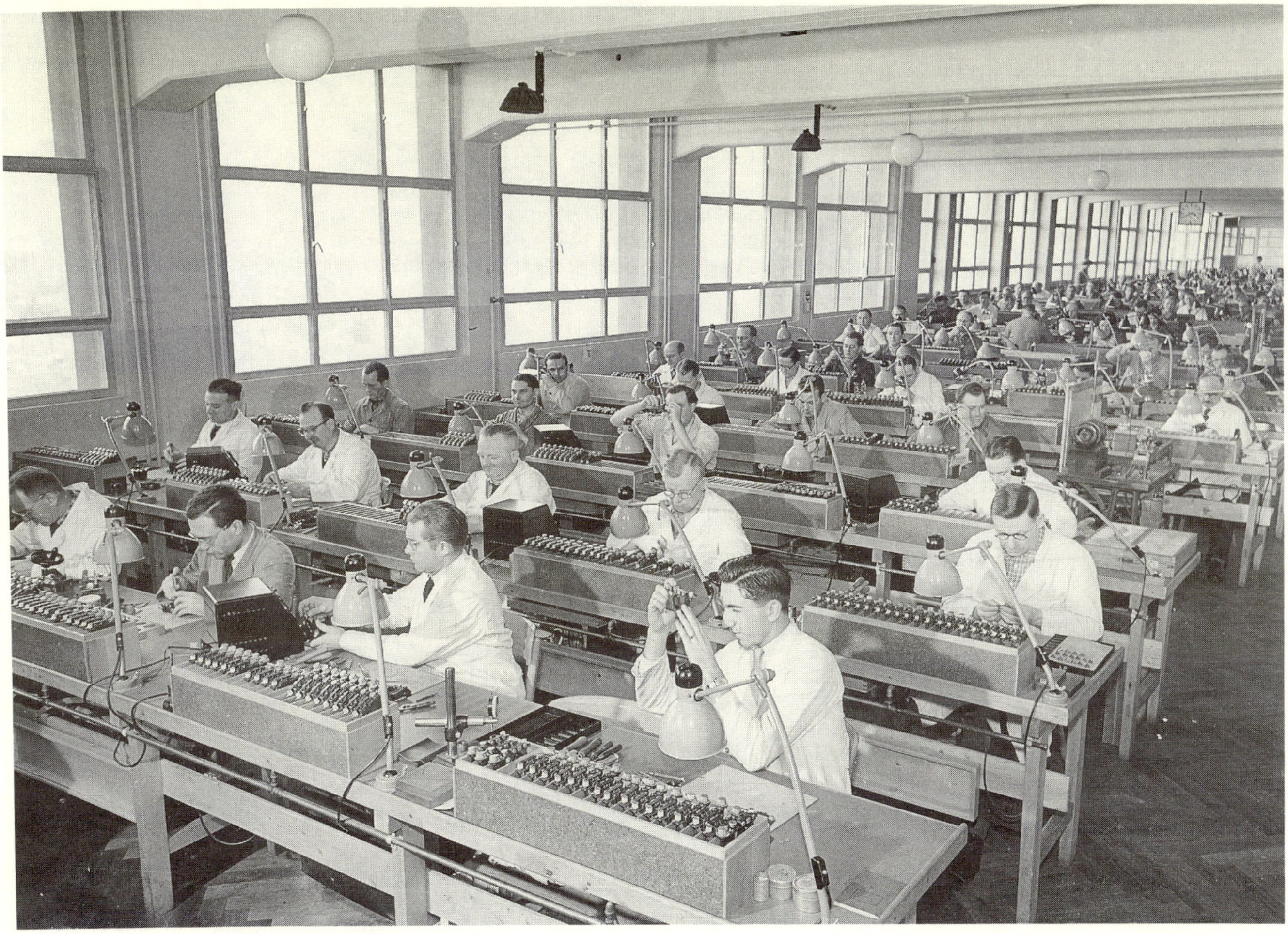

Leica production facilities at factory (1955)

books and advertising literature. Schwalberg had gained a great deal of experience in the photographic business, especially in Japan, and was an expert in product development trends.

While in Wetzlar, he continued to write articles for U.S. photographic magazines. His first visits to Germany were made with great trepedations, but he stayed for 1.5 years. Upon his return home he indicated to his superiors that the time he spent at Leitz Wetzlar had been the happiest of his career. He left with love in his heart for the place and its people.

When Leitz needed more working capital, they usually succeeded in obtaining a loan from the"Hessische Landesbank". Unfortunately, the photographic industry in Germany was at that time less than prosperous and loan commitments could no longer be extended.

Leitz Wetzlar was turned down and Henry Mann, the President of Leitz New York, became the first shareholder in 1962 in what had been to that point strictly a family business.

These ten years (1952 - 1962) were filled with many pleasant memories of a rejuvenated agency in New York. Undoubtedly , Mr. Mann enjoyed his role as president and shareholder, coming into contact with some of the most prolific and fascinating people in the sciences and photo-journalism.

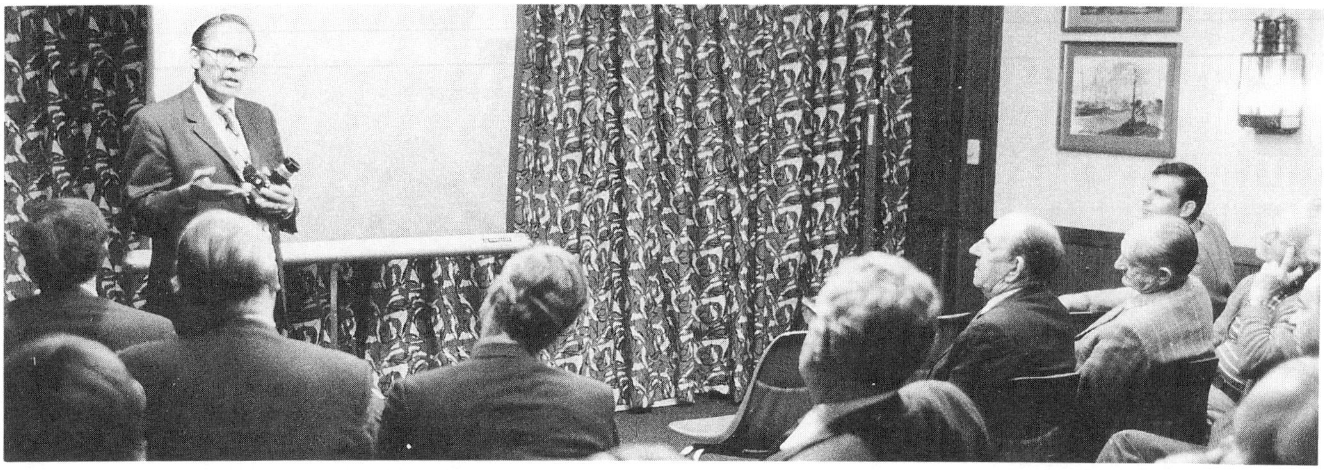

Walter Heun lectures "Leica"

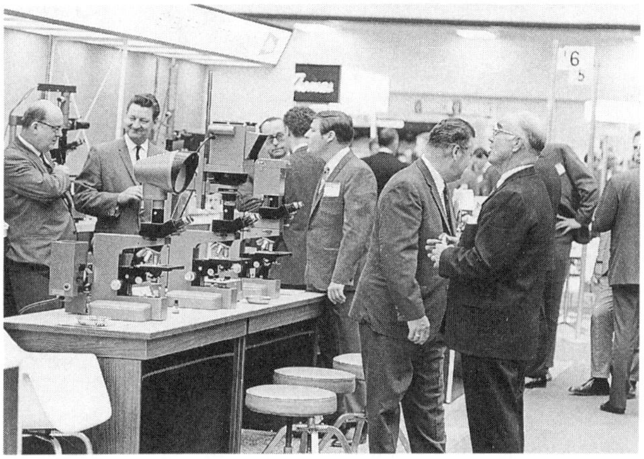

Leitz New York exhibits scientific instruments

Guenther Leitz with Dr. L. Beebe, recipient of a Leica camera with significant serial number, and Henry Mann watching

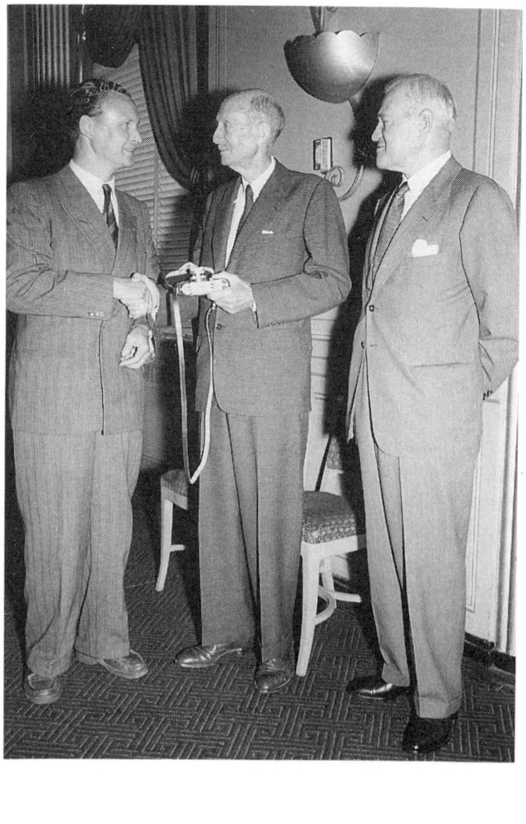

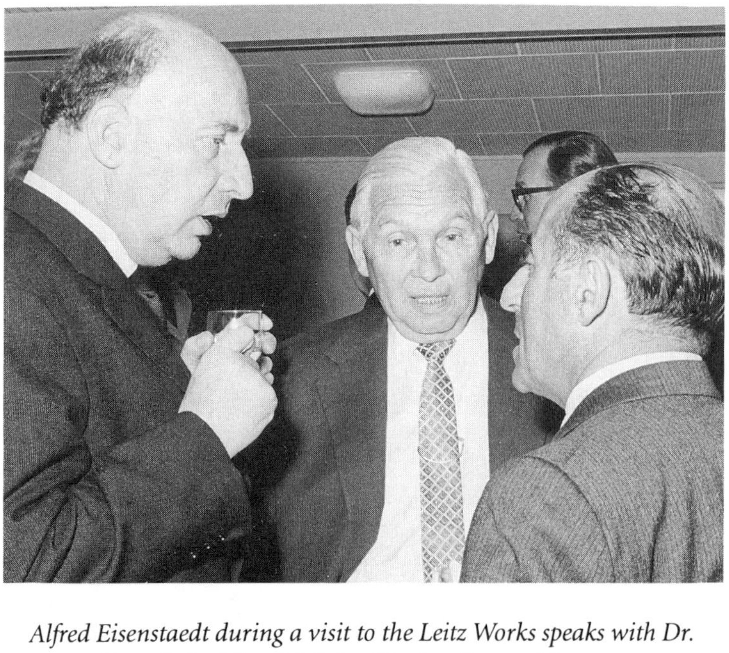

Alfred Eisenstaedt during a visit to the Leitz Works speaks with Dr. Ernst Leitz III and Michael Becker, Export Manager

Guenther Leitz, now in Midland, his new North American plant , came to New York and presented a Leica camera to Dr. R. Beebe for his outstanding achievements.

In 1957, the oldest daughter of Ernst Leitz II, Dr. Elsie Kuehn-Leitz, travelled throughout the U.S.A., visiting 28 states, gathering impressions which she conveyed to Konrad Adenauer, post-war president of Germany, during a visit on January 6, 1958, a day after his 82nd birthday. The life story of Dr. Elsie Kuehn-Leitz is found in a book by Klaus Otto Nass, her son-in-law, entitled 'Elsie Kuehn- Leitz Mut zur Menschlichkeit' published in 1994 by Europa Union Verlag GmbH, Bonn.

Elsie successfully pursued four major goals:

1. Re-establish culture in her home town of Wetzlar after the devastation and emptiness following World War II

2. Become involved in the plight of Africa and communicate with Albert Schweitzer

3. Establish contact with Konrad Adenauer and work for a United Europe

4. Rapprochement between France and Germany, exchange visits between French and German cities.

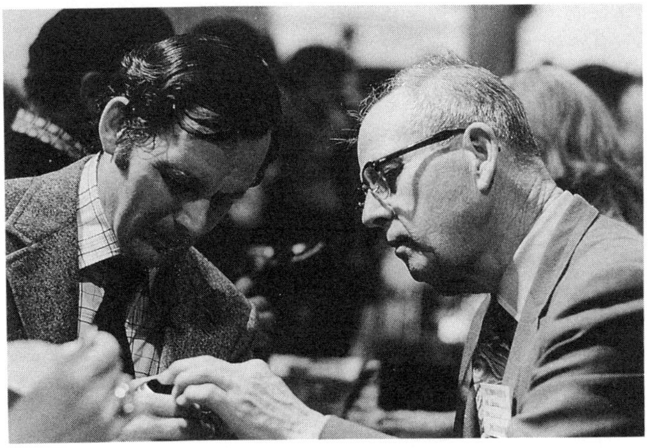

John Brooks, Leica (U.S.A.) School makes a point

A few years later, in 1959, Guenther Leitz returned to Wetzlar to become involved in international marketing in the capacity of his management position at the home factory. His understanding of world markets was strengthened by his experience in Canada, as well as the United States, which he visited frequently.

Guenther encouraged contacts with the world wide community of photo journalists as well as outstanding scientists. His enthusiasm and friendly personality sparked the work force to achieve higher and higher goals. But fateful tragedy would enter his life. After a brief illness, at age 55, he passed away.

Mr. Mann also had an employee who in 1964 had an anniversary of 50 years of service to E. Leitz, Inc., New York. Her name was Etta Hurvich ("Sugar" for short) who was hired in 1914 by Alfred Traeger. "Sugar" was always in the order processing department. She knew the price for every item in the vast catalogue of Leitz products. She also calculated the discount applicable to every item for dealer sales and added whatever tax had to be collected. The calculator and computer were in her head and she never forgot to credit the salesmen's commission to the right person. "Sugar" retired in 1964 and Henry Mann, when giving a luncheon in her honor, presented her with a watch.

The Chairman of the Board at Leitz Wetzlar, Dr. Henri Dumur, was 77 years old in 1963 and said: "I have been with Leitz for fifty eight and a half years. It would be high time to retire but my successor has not yet been found, so they say! It seems to me that they are not conducting a proper search!" The underlying mood of his comment was clear: it was time for a change. This interpretation represents the writer's view but the 6,500 people employed at the time in Leitz factories and offices were still operating under the leadership of a family enterprise which found it difficult to finance its operations entirely out of its own resources.

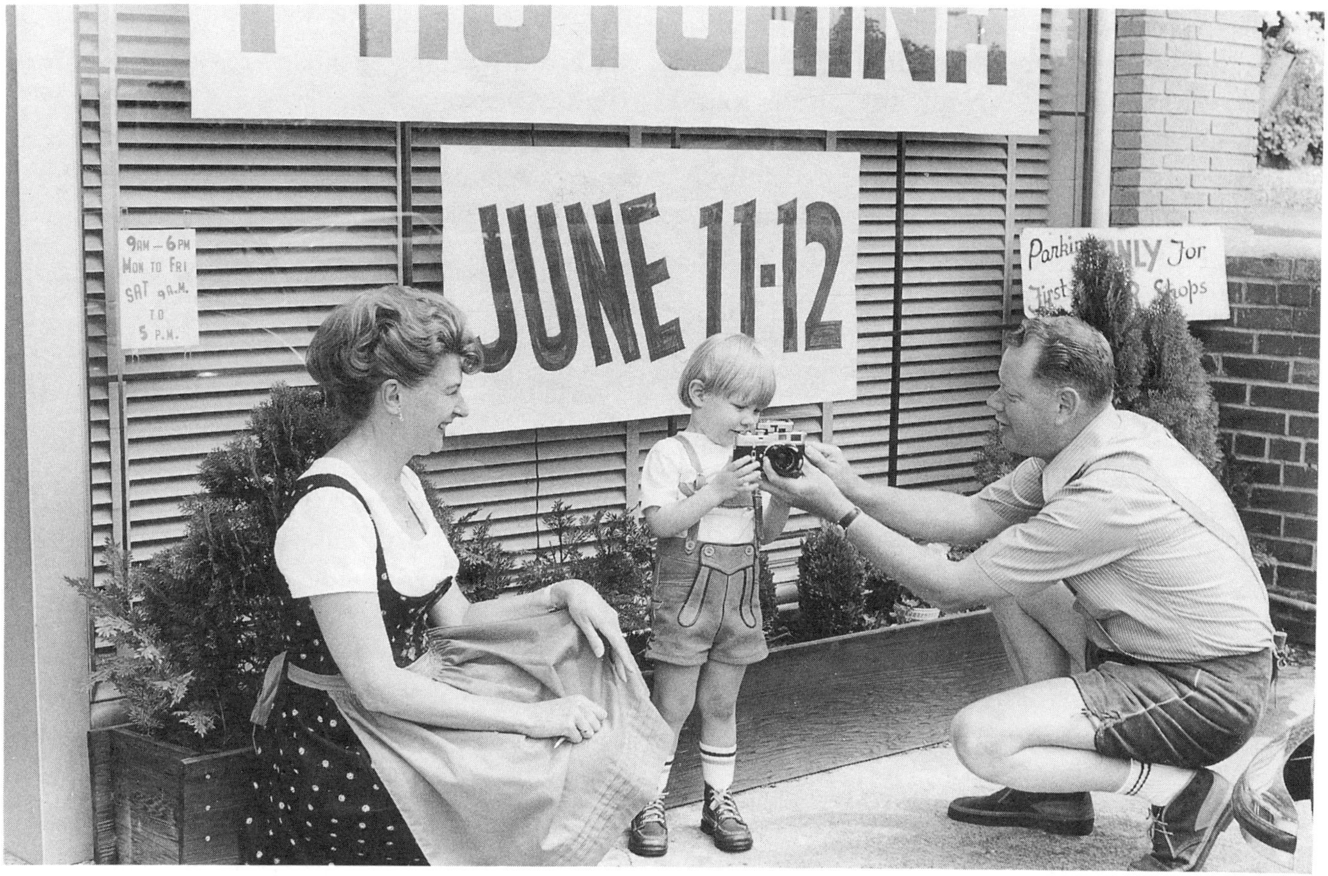

Steve Kippert gives a Leica demonstration to an up and coming young photographer

Chapter 10

HENRY MANN'S SECOND 10 YEARS

The Second 10 Years of Henry Mann's Ownership of E. Leitz, Inc., New York, 1962-1972

Before World War II, Leitz Wetzlar began to design and develop a new series of precision measuring instruments. In 1928, a profile projector was designed to control the accuracy of small parts such as gears. The impetus for this development came from Ernst Leitz III, and was intended not only to advance the mass production parts for the Leica, but also to expand technology in general. Rather than to integrate the distribution of these precision measuring tools, workshop measuring microscopes WM, etc. with its own established microscope and camera business, Leitz Wetzlar found a specialist in the U.S.A. with many years of experience in this field.

They selected the George Scherr Company, located at Lafayette Street, New York City. George Scherr successfully introduced this new product line to the American market. After the war, Scherr decided to retire and "Optometric Tools" was founded with F. Koenig and Adam Crystle in charge, both long time trusted employees of the George Scherr Company. Optometric Tools was a wholly-owned subsidiary of E. Leitz, Inc. They moved into the building occupied by E. Leitz, Inc., New York, at 304 Hudson Street, in 1950. In 1970 they moved in with E. Leitz, Inc., in Rockleigh, New Jersey.

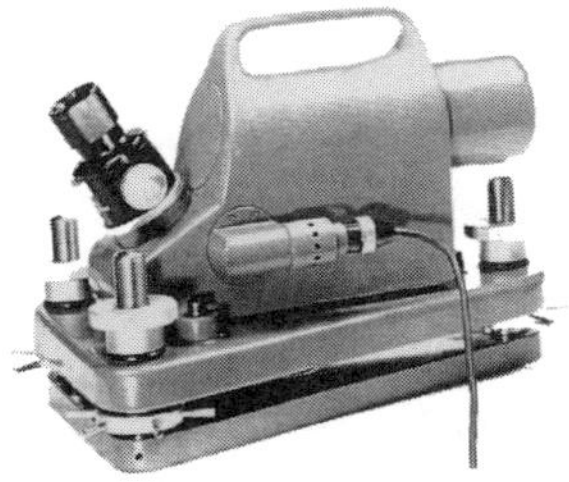

Auto-Collimator

In the field of metrology, the measuring microscope is an instrument used to measure geometric dimensions of parts or features by optical means. An object to be measured is magnified so that its geometric characteristics can be inspected and measured without physical contact. It is critical that the optics of such instruments do not create undesired aberrations. Following the profile projector, a workshop measuring microscope was developed with incident and transmitted light facilities to measure lengths and angles. Alignment telescopes followed.

By 1949, a comparator with perflectometer made headlines. It was the first length measuring instrument of high accuracy, complying with the Abbe principle. Then came the Master Dividing Head, the Auto-Collimator, a Light Sectioning Microscope for testing pitch, depth and profile, the Miniload Hardness Tester, and the Latimet to measure line widths of micro-

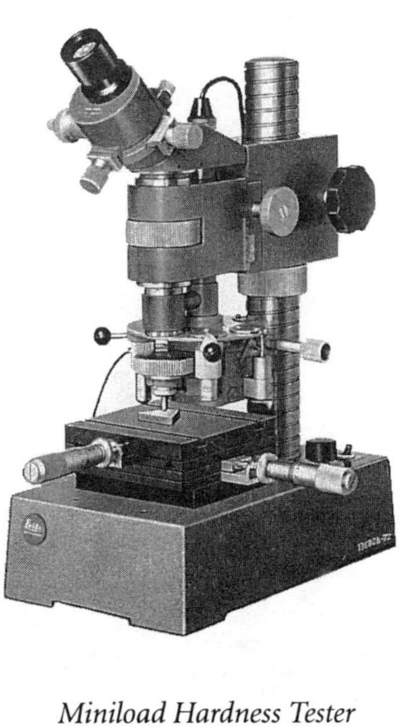

Miniload Hardness Tester

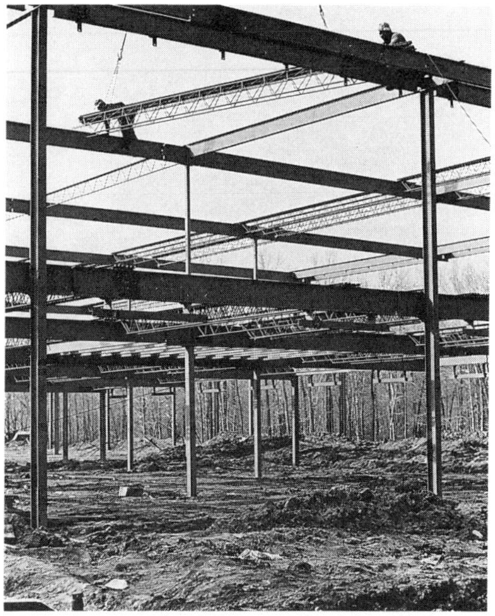

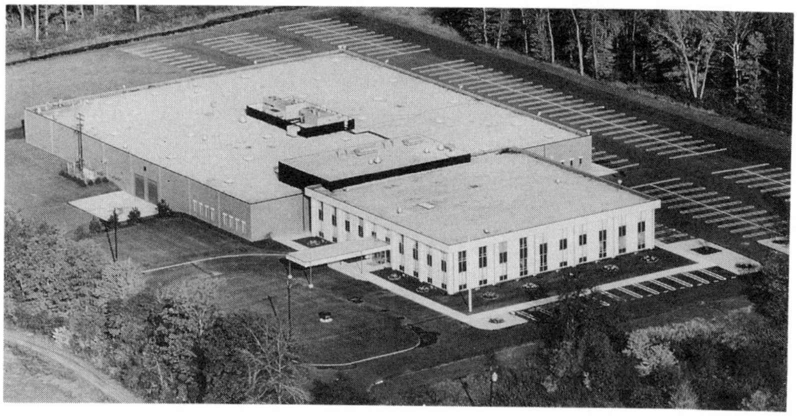

Leitz Rockleigh, New Jersey (1970)

Construction of new office, service and distribution facilities in Rockleigh, New Jersey (1968)

Arthur Koch, metrology production head, Leitz Wetzler

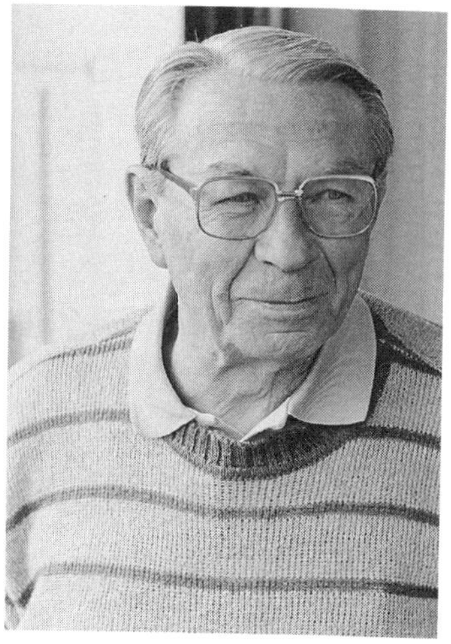

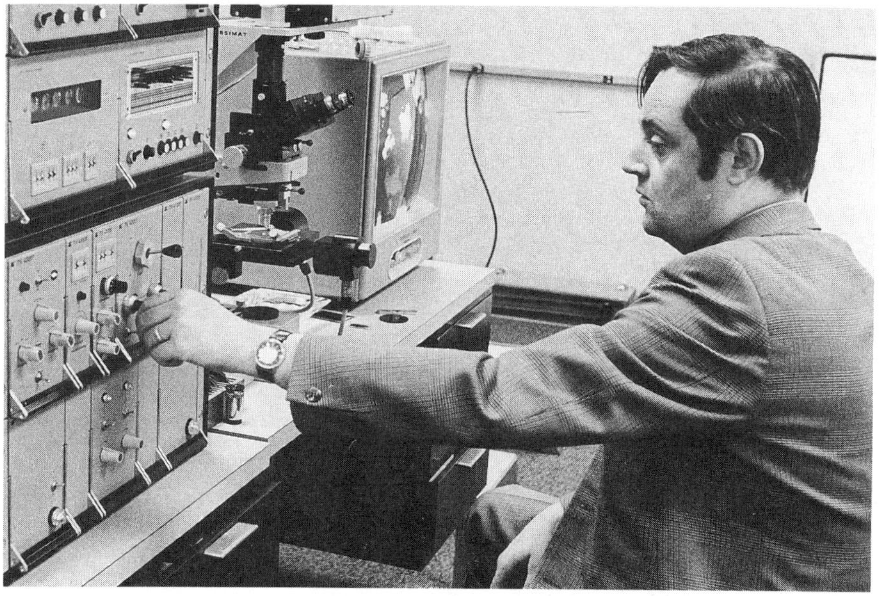

Werner Hunn at the controls of a computer-assisted Texture Analyser

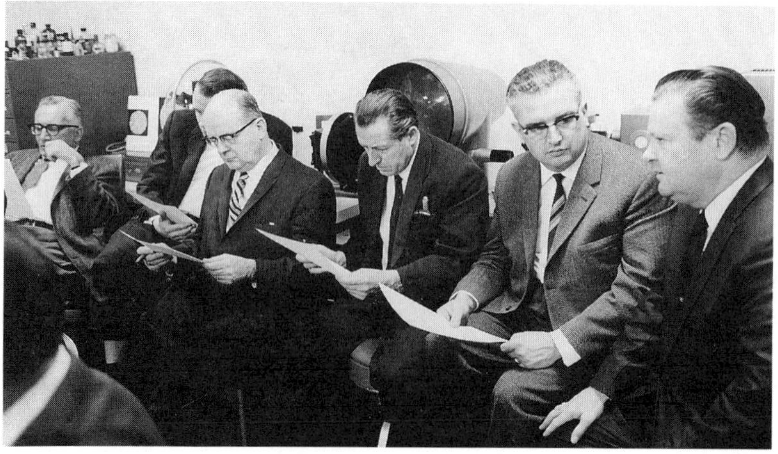

Leitz engineer Dr. Walter and Fritz Heckman (second and third from right) during Leitz New York sales meeting

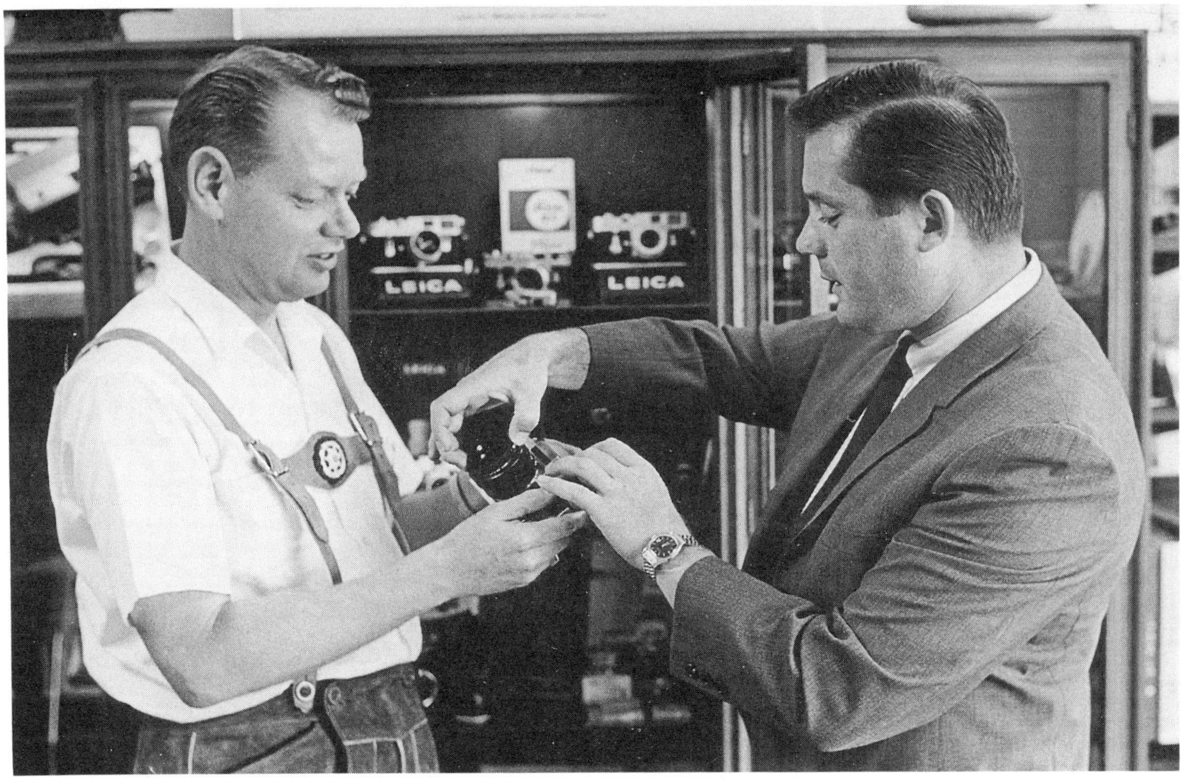

Steve Kippert and Mr. Wallis during Leica demonstration in Atlanta, Georgia

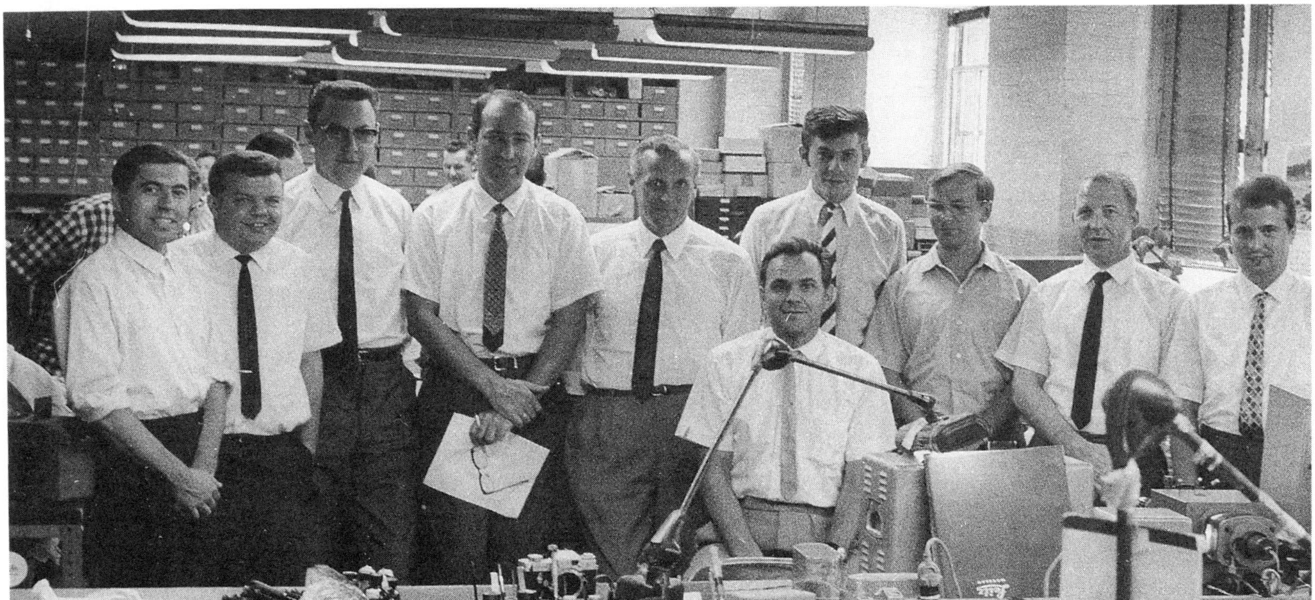

Leica repair training team for the U.S. Signal Corps (Friedrich Petri, Klaus Wiegmeyer, Manfred Neul, Helmut Krueger, Werner Keiner, Adolf Willhauer, Bruno Israel, Hans Steinbrecher, Ulrich Fuchs, Center: Otto Schmidt)

structures in semiconductor applications. In 1975, the "Correvit" was introduced for use in the automotive industry.

By 1960, total sales of E. Leitz, Inc., New York, reached $ 7,602,462, compared to just over $ 4 million when Henry Mann took over. Leica cameras and lenses sales were $ 3,900,000, microscopes, colorimeters and repairs $ 3,600,000.

By 1965, this increased to $ 4,500,000 for photo products and $ 6,500,000 for scientific instrumentation. The instrument design team at Leitz Wetzlar and the scientific community in the U.S.A. cooperated extensively with great success. Designers like F. Heckmann and others spent many weeks in the U.S.A. to study current research requirements and competitive activities, using their insights in the product development for emerging technologies.

A 1969 example was the importance of "recognition and automatic analysis of cell images" in cytology. Computer programs were designed to classify different cell types in the human body, classify them and determine automatically by comparison whether a specific cell falls into that category or whether it represents a new and different cell type.

Computer technology and microscope images together formed a new way to look at pathology. Self-learning computer programs gave rise to microscope/computer combinations with the promise to simplify and improve the accuracy of medical diagnosis. A new opportunity for medical research was at hand and the factory engineers were quick to respond. An automated stage was added to industrial research microscopes to scan metallurgical samples with a computer-assisted texture analyzing system. Focusing mechanisms became automated.

Meanwhile, in the photo products line, sales for the Leicina motion picture camera were diminishing. But, another long overdue photo product, the Leicaflex, was introduced. A Navy contract for cameras and lenses was received from the Signal Corps, which purchased 2,000 IIIf cameras with 35/50 and 135 mm lenses. By 1966 , both disciplines showed substantial sales growth. Total sales now amounted to $ 12,143,991 and more than $ 7,200,000 of that total was attributable to instrument sales. Between 1967 and 1969, Central Naval Procurement in Washington D.C. purchased 480 - 500 Leica M2-R cameras with 35/50 and 90 mm lenses for distribution to its naval reporters. Again in 1969, the Navy purchased 200 Leicaflex kits with lenses from 21 to 400 mm focal lengths. Each kit was assigned to an individual reporter.

This continued growth sparked the planning of new and expanded facilities once again. It followed the trend of the times, which was to move out of the urban centers, such as New York City, where room for expansion was limited. Most of the employees favored the idea of getting away from asphalt and concrete to the green lawns, fresh air and serenity of the suburbs.

Rockleigh Industrial Park was the selected site, located in the northern part of New Jersey, still relatively close to New York City. It was a mere 30 minutes drive from the George Washington Bridge. The company now employed more than 250 persons, plus 25 sales representatives in the field. Twenty-two of these were repair and service mechanics for photo products and 5 for scientific instrumentation. All in all, the new building provided an 86,000 sq ft. operating area with the most modern loading and unloading facilities to efficiently handle container air shipments from the factory. Leica schools and microscope laboratories were given new prominence and everyone seemed to sense a new and prosperous future.

The dedication by William Mann , who was the President of the company at that time, took place on April 28, 1970. The disadvantages of having two Manhattan locations - one in

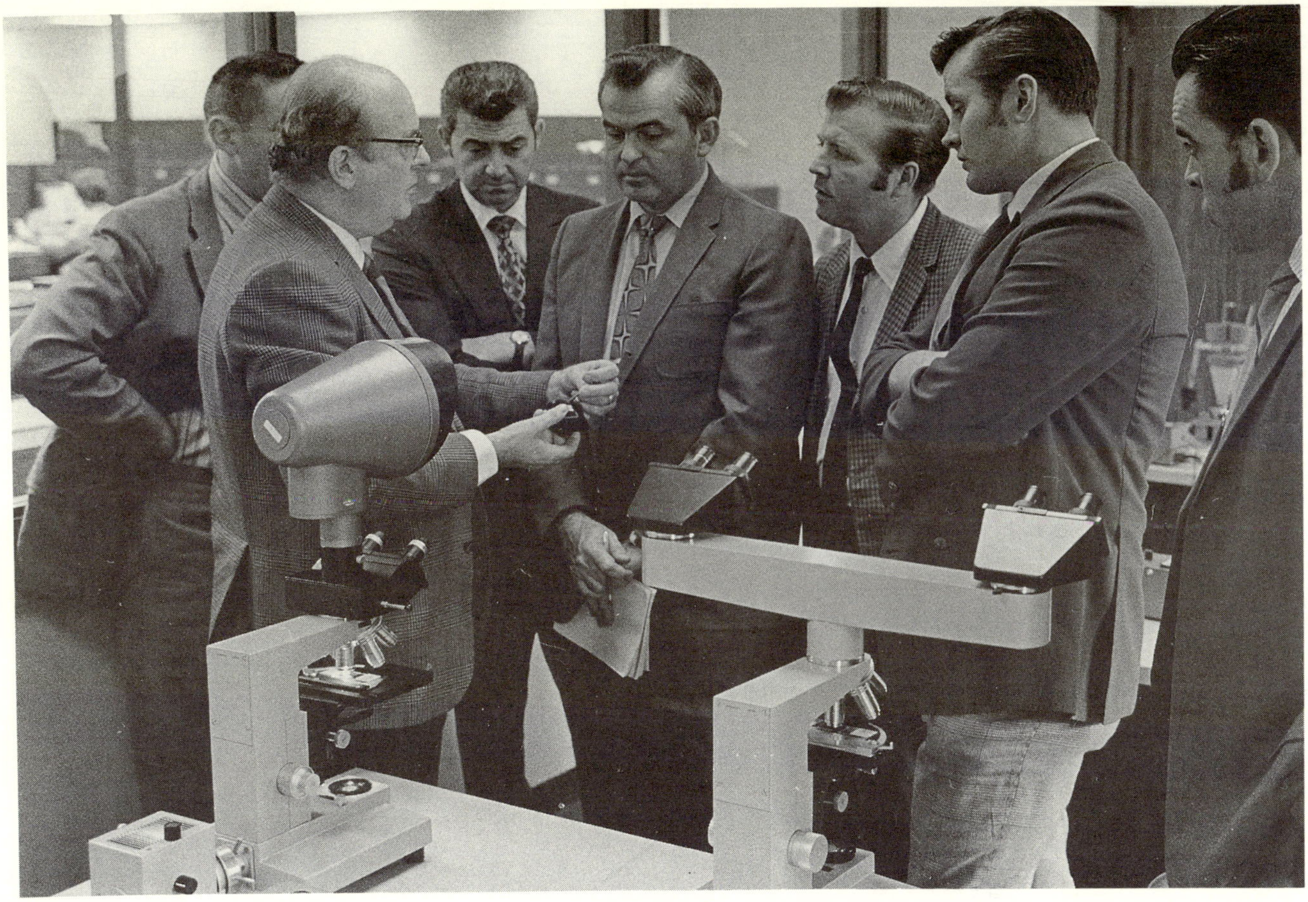

Rudi Kahn demonstrates Leitz comparison microscopes to the Crime Laboratory of the New York Police Department

midtown for administration and sales, the other four miles away in lower Manhattan for service, manufacturing, repair, inventory and shipping had been overcome.

Unfortunately, Henry Mann was never to see the new premises, as he suddenly passed away in August of 1968. He had lived a full life. Born in 1890, an immigrant to the U.S.A. from Germany in 1913, he rose to become controller of the Crown, Cork & Seal Company in Baltimore. He was a Director of the American Chamber of Commerce in Berlin and settled in New York after the outbreak of World War II, when he formed his own securities company on Wall Street. He was buried near his East Hampton home. Guenther Leitz presented the eulogy at the grave site.

In the years to follow (1972-73) new photographic products boosted sales further. The unique and sophisticated Leica M5 was introduced. Willi Stein at the same time completed the design of the so-called "People's Leica", the CL model. This again revived the company's market share in the lower price range. The CL was readily accepted by a broad spectrum of camera enthusiasts who found its compactness, easy handling and light weight particularly attractive.

Microscopes were now becoming indispensable in the large semiconductor industry which showed enormous growth. Combined with photo products, Leitz New York now registered an annual sales total of $ 16,967,171.

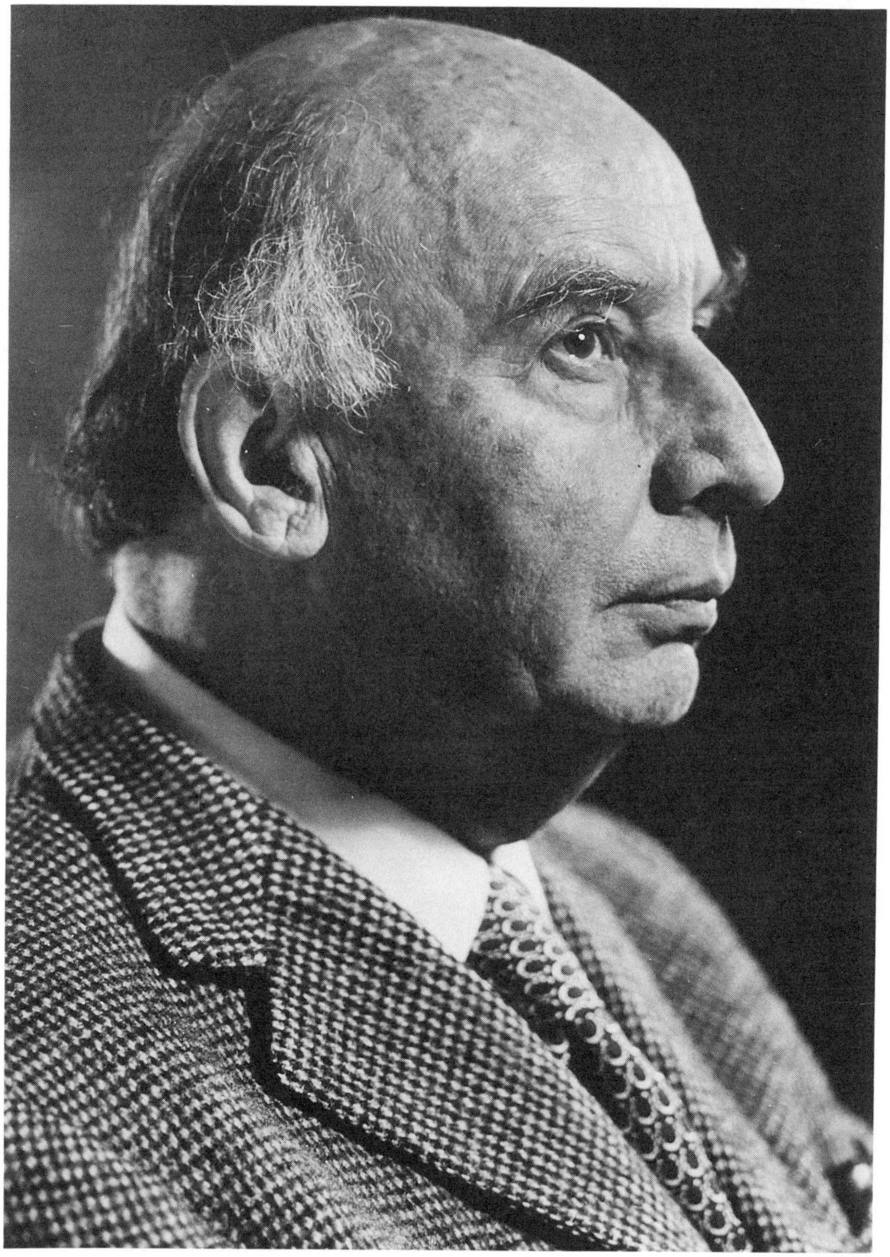

Dr. Ernst Leitz III

Chapter 11

THE NEW LEITZ HEADQUARTERS

The Dedication of the New Leitz Headquarters in Rockleigh, New Jersey, on April 28, 1970 Optometric Tools, Subsidiary of E. Leitz, Inc.

An invitation was sent to all Leitz/Leica friends in the New York metropolitan area to participate in the inauguration of the new Leitz headquarters in New Jersey. Presentations were made by Dr. Ernst Leitz III, the principal owner of the Leitz family enterprise in Wetzlar, Germany, and William H. Mann, the President and owner of the new Leitz U.S.A. headquarters.

After the reception, tours of the building were conducted, during which the employees proudly exhibited the new improved facilities for all aspects of the operation - from shipping and receiving, to service, repair and manufacturing, data processing, training schools, advertising and administration. This was their new home, fully air-conditioned with 86,000 square feet on seven acres of land.

Optometric Tools, Inc., a subsidiary of E. Leitz, Inc., was now included on the new premises. Its ultra-precision measuring microscopes, its precision optical dividing-head, used to guide Apollo on target to the moon,were on diplay. Additionally, rotary index tables, coordinate measuring stages, collimators, interferometers, alignment telescopes and toolmakers' microscopes were on display next to a glass enclosed vitrine, displaying the development stages of the Leica camera.

After the war , Leitz Wetzlar had been increasingly involved in metrology, becoming a major producer of measuring microscopes, serving an

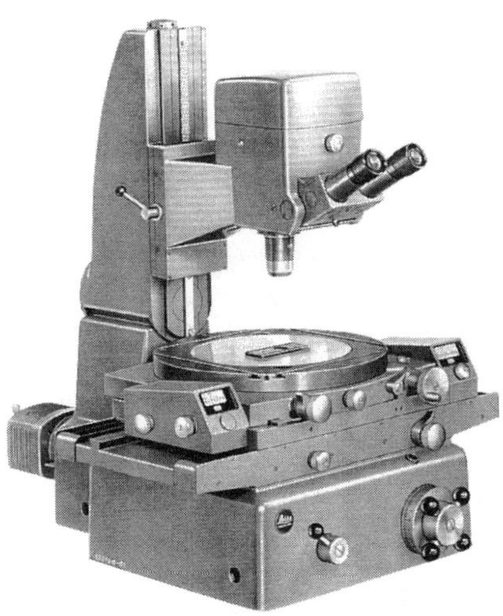

Large Toolmaker's Microscope

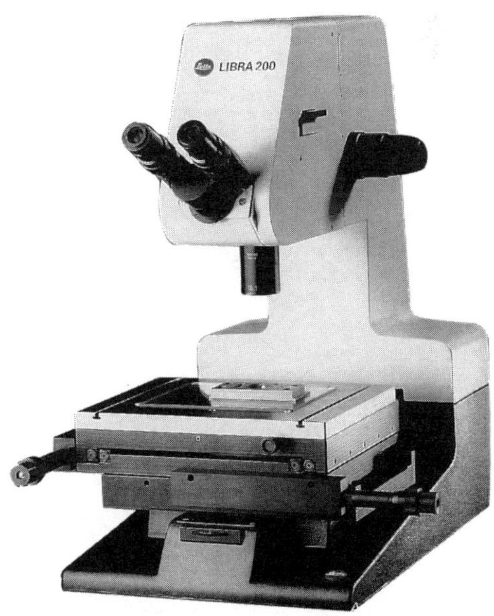

Libra 200, a new generation of Measuring Microscopes

important function in industrial applications. Such instruments range from simple component devices to fully automatic, computer-controlled universal measuring systems, together with the appropriate software. These are optically based, non-contact, dimensional measuring instruments that allow measurements which would be virtually impossible by direct contact. Leitz earned a world wide reputation in optics and their measuring microscopes became the standard of the industry. Some of the manufacturers which opted for Leitz measuring microscopes are: Aerospace Industries, Automobile Manufacturers, Consumer Products (razor blades), Electronic Industries, Semiconductor Manufacturing, etc. to name some of the important ones. Optometric Tools, as a division of E. Leitz, Inc., now distributed a full program of optical measuring systems, recognized for their outstanding measuring capabilities and flexibility together with proven accuracy and repeatability.

The largest of several display rooms featured the tremendous variety of Leitz microscopes and photomicrographic apparatus used in almost every scientific discipline from medicine to quality control of computer chips. It was a display of optical- and precision mechanical equipment second to none in the world.

The guests received a commemorative silver coin, the size of a silver dollar ,dated April 28, 1970. It depicted on one side Ernst Leitz GmbH, Optical Works, Wetzlar, with a Leicaflex/Microscope combination in relief and E. Leitz, Inc., Rockleigh, N.J., on the other. Mr. William Mann used the occasion to congratulate his employees for their loyalty and dedication.

A few years later, in October 1975, the 50th anniversary of the Leica camera was celebrated, both during the annual meeting of the Leica Historical Society in Dearborn, Michigan, as well as in Rockleigh, N.J. the headquarters of E. Leitz, Inc., U.S.A. and, of course, Wetzlar, Germany. Dr. Ernst Leitz III visited on that occasion and presented the principal address before the LHSA. He recalled the history of Leica's development - a theme that he said was "not a stranger to this audience"! It would be redundant to review his complete message, as it was fully reported in Viewfinder Volume 8, No. 4, October/December 1975. Dr. Leitz did not hesitate to cite his own contributions to the development of early Leica accessories as well as the introduction of lens grinding methods, using diamond tools. An exchange of views followed and the lively audience was enchanted by the unique occasion.

The 50th anniversary party held at E. Leitz, Inc., Rockleigh, included eminent representatives of the photographic press as well as the founder of "Photokina", Dr. Fritz Gruber. Both, Mr. William Mann, President of E. Leitz, Inc., as well as Dr. Ernst Leitz III underscored Leica's path of "growing up" and predicted a positive passage for its future. Rolf Fricke, an early LHSA member, loaned his Leica collection for the occasion and, more than anything, he provided the background for the discussions that followed the viewing of these outstanding examples of Leica lore.

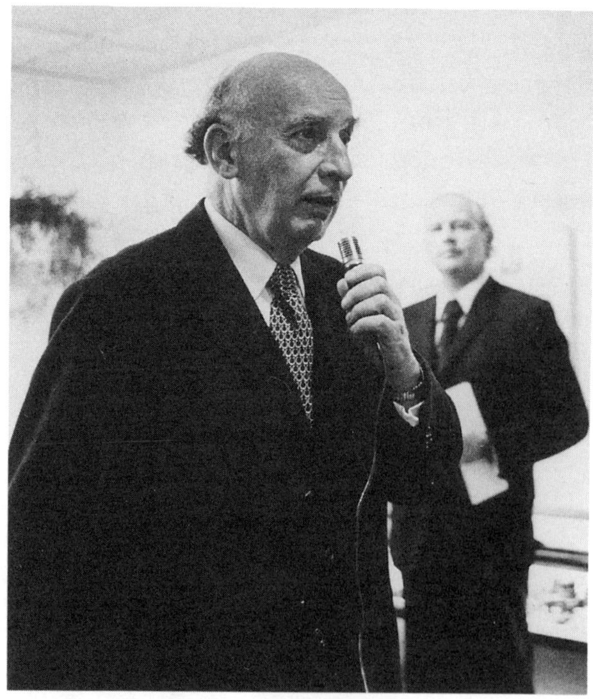

Ernst Leitz III speaking in Rockleigh, NJ during Leica's 50th Anniversary, with William H. Mann in the background

The wind of change was sweeping once more, this time with respect to economic conditions, both here and abroad. Skilled toolmakers, which are indispensable in this industry, earned more than three times the wages in Germany when compared to their counterparts in Japan. The U.S. employment demographics changed in the 1970's from industrial productivity to service jobs. The balance of trade started to show gains by foreign nations so that it became necessary for the U.S.A. to impose an import surcharge vis a vis goods from the European community, Japan and even Canada.

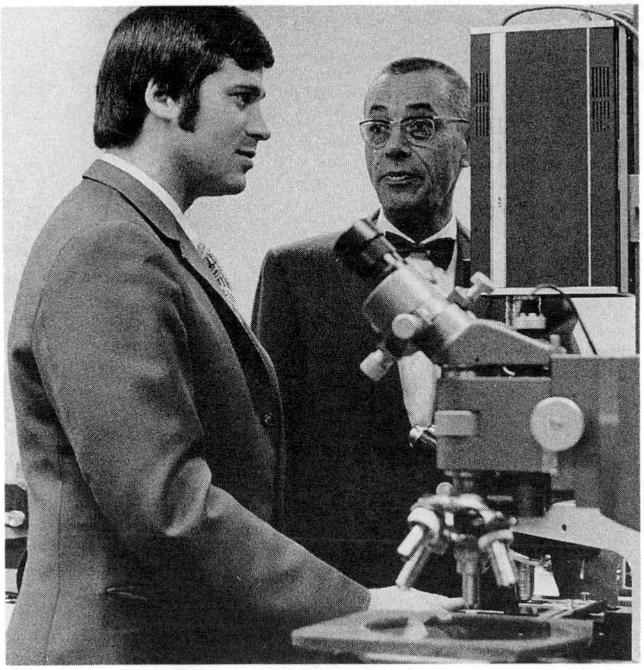

Horst van Schellenbeck demonstrates instruments distributed by Optometric Tools

Leitz, in order to stay competitive in the world market, established a branch factory in Portugal where labor was less expensive. Cooperation was now also sought with Leitz' strongest rivals in Japan to produce the Leica "CL"cost effectively. Ultimately, by 1974, it became necessary to sell 51%, which constitutes the controlling interest of Ernst Leitz Wetzlar GmbH, to Wild Heerbrugg/Switzerland. The impetus for this sale had been given in the late 1960s by Henry Mann. The merger with Wild was intended to reduce the duplicated scientific instrument product lines of the two firms.

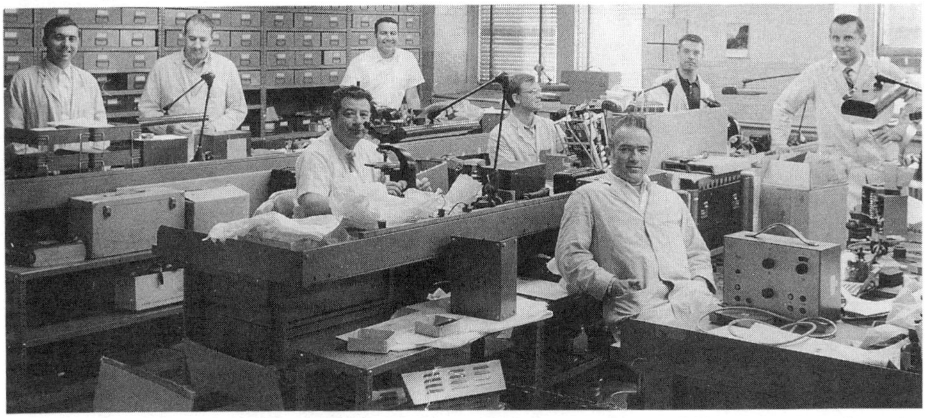

Microscope repair and service department personnel

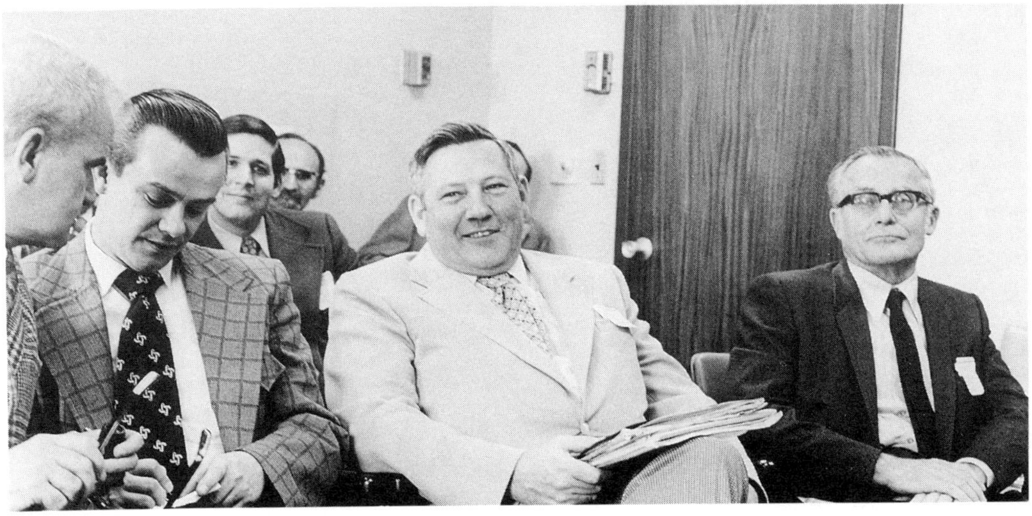

Microscope dealers attend Leitz Microscope School at Leitz Rockleigh

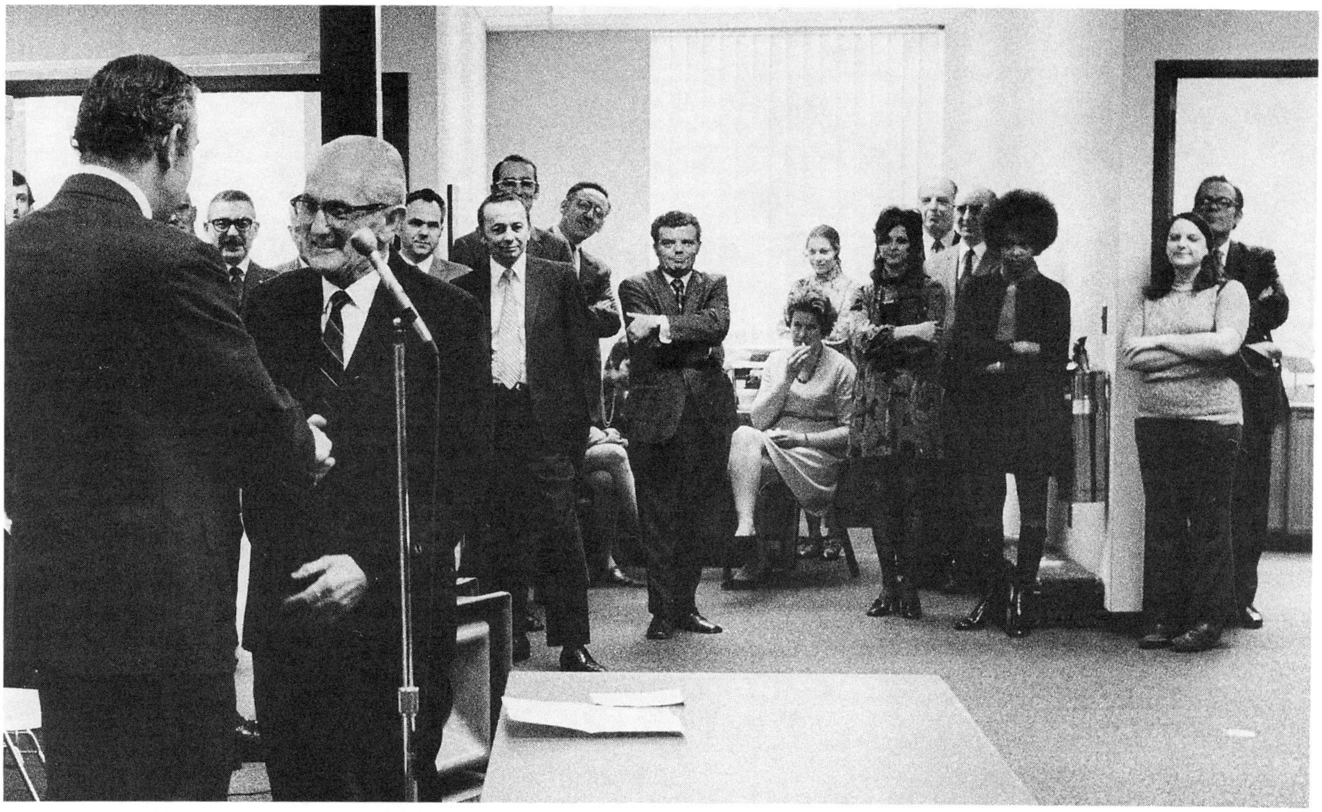

Leitz New York, employees are recognized by William Mann, President, for 25 years of loyal and successful work

Chapter 12

25 YEARS OF MANN MANAGEMENT

25 Years of Mann Management
- September 28, 1977 -

This was a time for reflection, of pleasure, recollecting accomplishment, thankfulness and of crediting almost 270 employees with unparallelled achievements. In 1972, the Agency Agreement between Leitz Wetzlar and William Mann was renewed for another 10 years.

During his presentation, President William Mann recalled the takeover of the company in 1952 by his father. He recalled the charted course from which the company never deviated, namely: "To represent the Leitz products in the U.S.A. in a manner commensurate with the high quality and integrity of the legendary Leitz name!" The special dedication of this day went to Alfred Boch, Vice President and General Manager, who retired after 57 years of Leitz company service here and abroad. To again quote from William Mann's speech: "Alfred Boch has set a standard of excellence which can only be described as magnificent, our commitment must be to emulate that high standard to assure that we will achieve further growth and success!" To convey his gratitude, Mr. Mann presented Alfred Boch with the keys to a new Mercedes 280E, which Boch accepted deeply moved.

Dr. Knut Kuehn-Leitz, representing the Leitz factory in Germany, recounted the acquisition of Leitz New York by the Henry Mann family. He recalled the history of Leitz U.S.A. since 1890, passing through two world wars, and twice under the administration of the Alien Property Custodian of the U.S. Department of Justice. After the takeover by the Henry Mann family, the company quickly realized its potential, in spite of the relatively difficult economic conditions of that time.

Dr. Knut Kuehn-Leitz mentioned the success of the Leica M3, introduced to the U.S.A. in the 1950's, the Leicaflex during the 1960's, and the growth of the market for medical and industrial research microscopes and other scientific instruments. To quote from his speech: "The late Henry Mann not only conducted his business in a professional way but also took great interest in Ernst Leitz Wetzlar. He became a friend of our family and it was his idea to bring the European optical manufacturers more closely together, ultimately resulting - after his death - in the cooperation between Leitz and Wild Heerbrugg, Switzerland." He stated: "William Mann had to take over the responsibilities for E. Leitz, Inc. during a period of great change. Today, I am pleased to state economic problems appear to be overcome. The future increase in the penetration of the American market through Mr. Mann and his team is assured!"

The occasion was marked by the presence of several dignitaries and associates. Dr. Knut Kuehn-Leitz represented the Wetzlar factory, as did his mother, Dr. Elsie Kuehn-Leitz and Mr. and Mrs. Alfred Loew , Sales Management. Mr. and Mrs. Walter Kluck represented Canada and Mr. and Mrs. Perrot had come from Switzerland. Local guests included consultants to the Mann family,

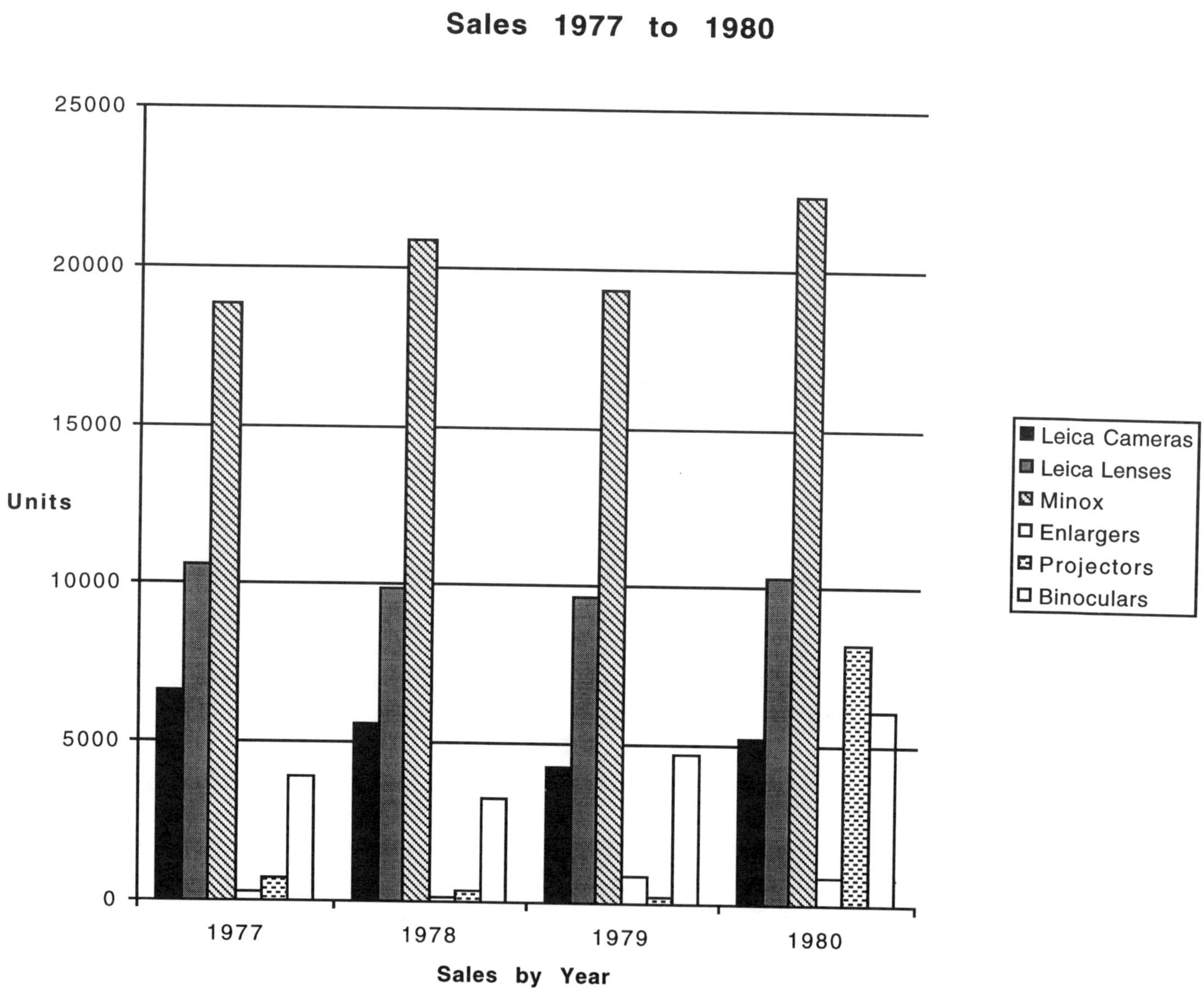
Sales 1977 to 1980
25000
20000
15000
10000
5000
0
Units
1977
1978
1979
1980
Sales by Year
Leica Cameras
Leica Lenses
Minox
Enlargers
Projectors
Binoculars

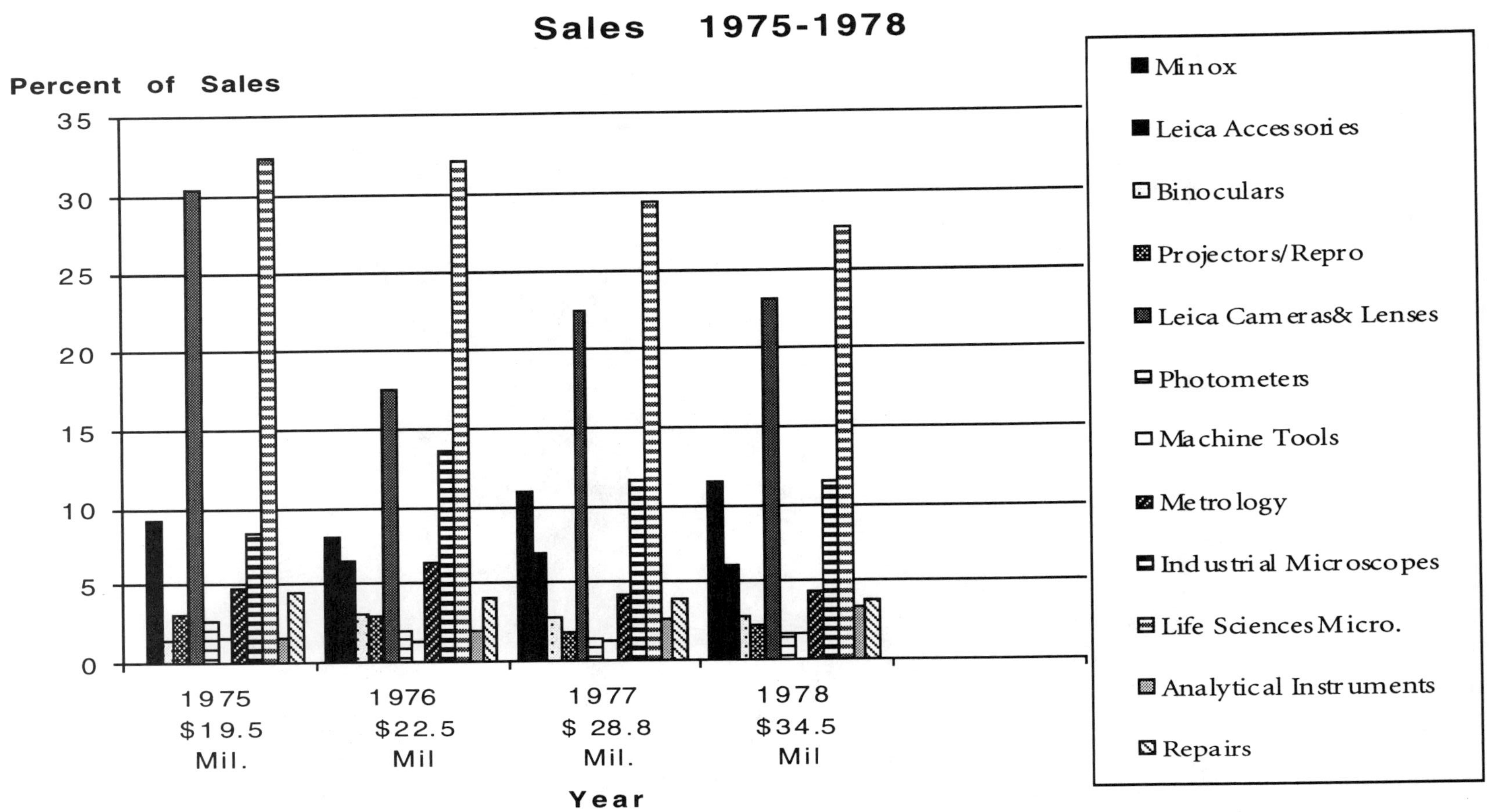
Sales 1975-1978
Percent of Sales
35
30
25
20
15
10
5
0
1975
$19.5
Mil.
1976
$22.5
Mil
1977
$ 28.8
Mil.
1978
$34.5
Mil
Year
Minox
Leica Accessories
Binoculars
Projectors/Repro
Leica Cameras& Lenses
Photometers
Machine Tools
Metrology
Industrial Microscopes
Life Sciences Micro.
Analytical Instruments
Repairs

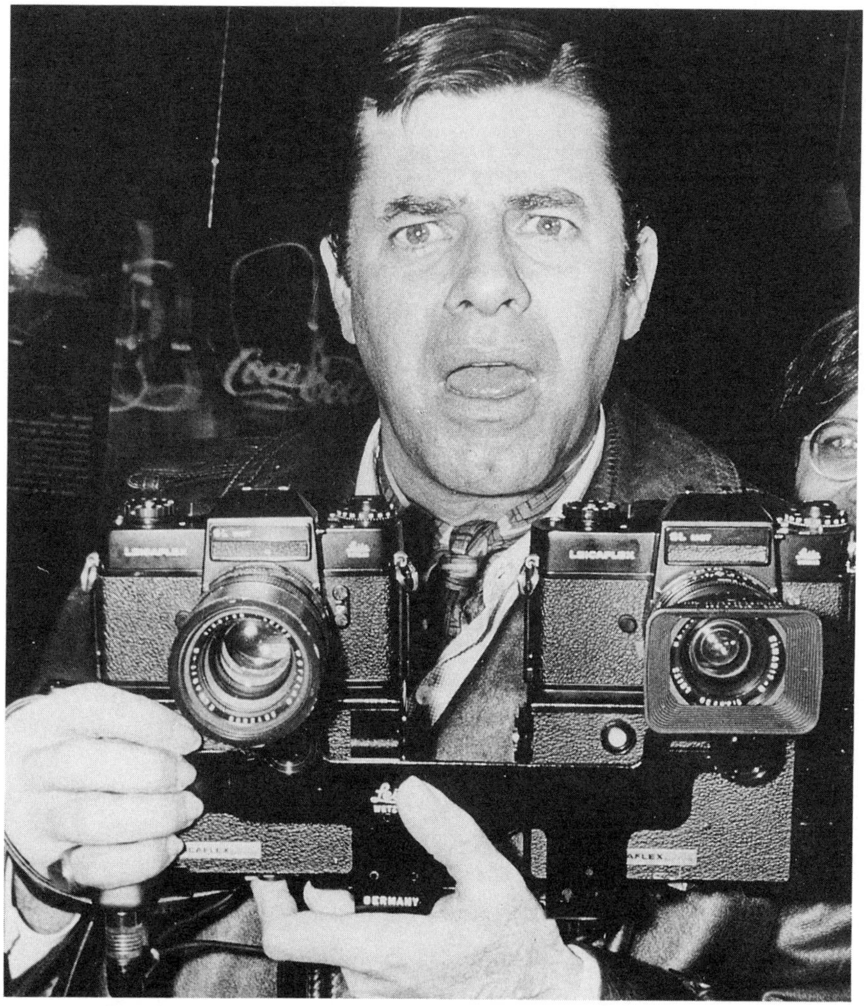

Dual Leicas astonish Jerry Lewis, the great comedian and Leica photographer

the law firm of Alexander & Green, Leitz New York's advertising agency, several banks, accountants, customs and forwarding firms and the Eastman Kodak Company.

This writer handed Mr. Mann a simple communique whose essential message was: "To follow the unique inheritance of tradition with progress, thus enabling all participants to celebrate this occasion and look with pride into the future."

These statistics record the sales of E. Leitz, Inc., New York, between 1977 and 1980:

a) For photographic products: Lcica cameras, Leica lenses, minox cameras, enlargers, projectors and binoculars.

b) For Biological Scientific Instruments: Student and laboratory microscopes, research and polarizing microscopes, photomicrographic cameras and microtomes.

c) For Industrial Microscopes: Inspection and metallurgical microscopes, micro-metallographs and hardness testers.

d) Metrology Instruments (Optometric Tools): Workshop precision measuring microscopes, multipurpose measuring instruments, profile projectors, mask comparators, optical dividing heads, IKL, ITP and Filar micrometers.

e) Analytical Instruments: Microscope Photometers, computer classified scanning systems.

f) Machine Tools: Imperial precision positioning tables and Fibrotakt tables.

In spite of relentless competitive pressures from Japanese and European manufacturers, E. Leitz, Inc., maintained its position as a major player in the diverse precision optical and photographic market of the U.S.A..

Regional offices were now in place in Tucker, Georgia, and Kensington, Maryland, representing the Eastern part of the U.S.A.. The North Central region was covered from Park Ridge, Illinois.The Southern regional office was located in Houston, Texas, the Western regional office in San Francisco. The Northeast was covered from Cincinnati, Ohio.

Thirty eight sales representatives from their respective regional offices were now in the field to carry the Leitz message to every corner of the United States. Quite a change from 1952, when no more than 5 or 6 salesmen tried to reach the market, both for Leica cameras and Leitz microscopes.

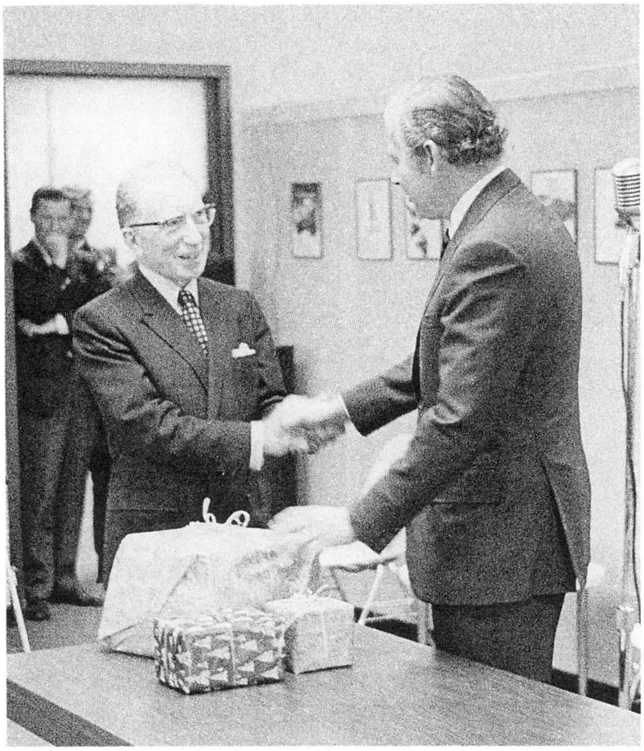

Special thanks to George McCaffrey, friend of the father and administrative executive for 25 years, extended by William Mann

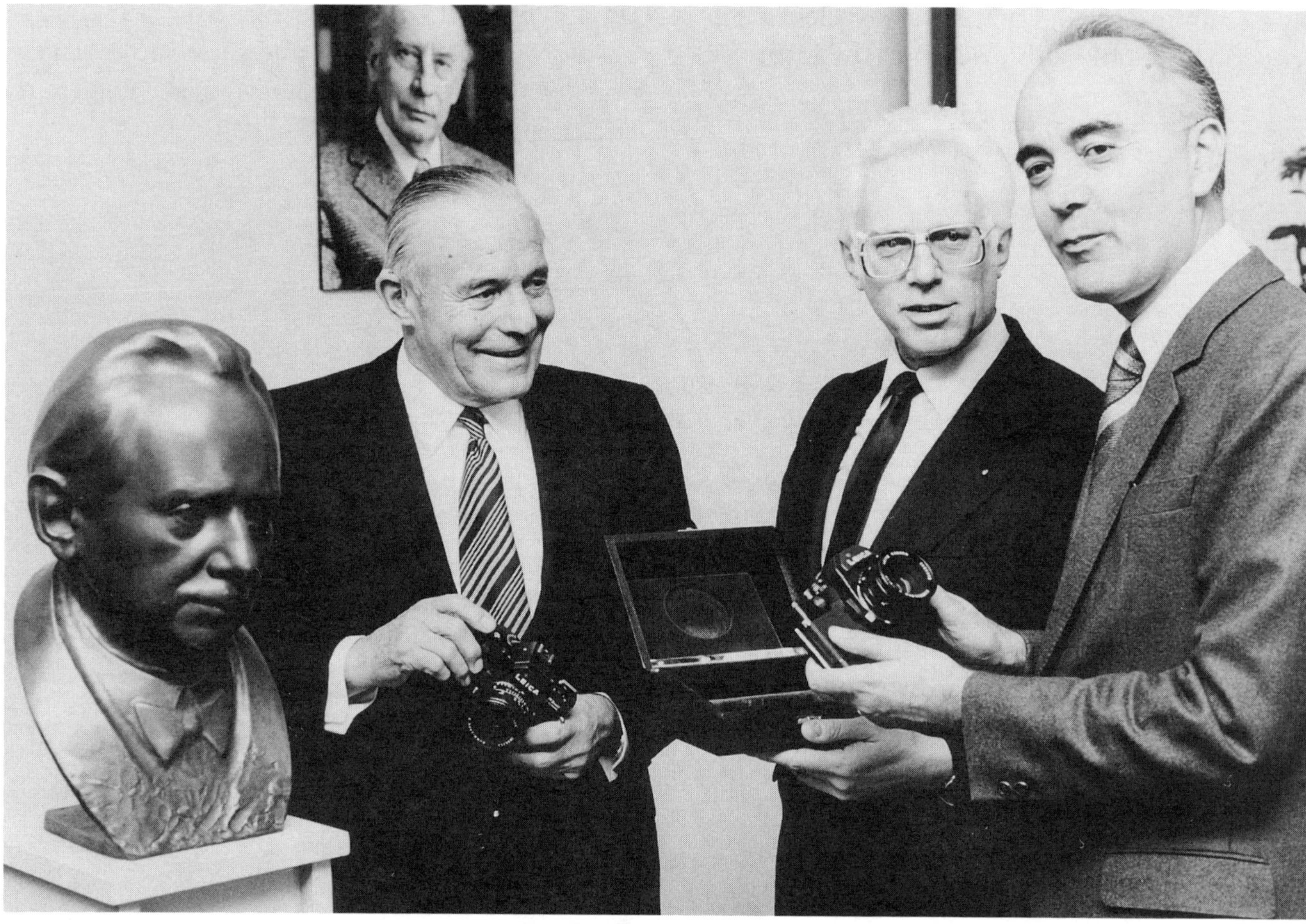

Dr. Hans Friderichs (right), President of the German Society of Photography received Leicaflex R3 # 1,500,000 while Dr. Max Kreis, Administrative Council of Wild Heerbrugg A.G. and Chairman of the Company Council of Ernst Leitz Wetzlar GmbH accepts # 1,500,001. The presentation was made by Alfred Loew, Marketing Management.

Chapter 13

STRATEGIES FOR THE 1970'S

Leitz Wetzlar Strategies for the 1970's

A 1971 review by Leitz Wetzlar of worldwide financial, marketing and competitive conditions, concluded that changes were necessary if Leitz was to continue its pre-eminent standing and market share in the global opto-mechanical and photographic industries.

Total sales amounted to 170 million DM; employment had reached 6,000 and competition from Japan had cast its shadow for some years. More than 54% of all Leitz products (microscopes, Leica cameras, industrial measuring instruments, etc.) were exported. The Leitz family owned 91% and Leitz New York, William Mann owned 9%.

Competitors from the Far East were not only gaining markets in the photographic sector, they also successfully penetrated the student and laboratory microscope market. High labor costs and costly union wage agreements plagued Leitz's competitiveness against countries with considerably lower wages. Public funds for research and development were diminishing.

Well placed sources indicate that the Leitz Works in Wetzlar made a critical decision in 1955 to change from private family ownership to an A.G. or "Aktiengesellschaft", taking the company public.

Venture capital was being used for the new administration building, a very expensive undertaking. To fund new production projects, capital must be committed many years in advance on a perpetual basis, each new project receiving the appropriate funding. Basic research should be funded at least 10 years ahead of production. The Leitz works did not make such commitments and suffered the consequences.

When applying for a loan at the "Hessische Landesbank" Leitz was denied. Several other camera firms, such as Franke and Heidecke, were equally in need of funds for capital development at the same time.

This condition forced management to consider alternative solutions. Its position in the microscopy sector was strong, offering outstanding instrumentation for all sectors of science and industry. Production of electronic hardware and software manufacturing had started. The photo sector, however, lost market share and the bottom line at the end of 1971 was negative for the first time in its history; and working capital was small.

Attempting to maintain its independence, Leitz needed working capital for production. A minority partner was necessary. Secondly, to maintain its position in the photographic sector, Leitz entered into cooperation with a Japanese competitor and established a manufacturing facility in Portugal, a low wage country.

To secure its position in the microscope market against strong Japanese competition, cooperation with WildHeerbrugg/Switzerland was negotiated. This arrangement promised synergies in product lines as well as the strengthening of the worldwide distribution channels.

Equally important was the strong capital resource of the Schmidheiny family . By 1972, Wild became a partner to Leitz with 25% participation. Its "Sperrminoritaet" (minority veto rights)and its ability to countermand executive decisions, gave Wild a powerful voice in the conduct of the Leitz business. Dr. Max Kreis became Chairman of the Board, followed , upon his retirement, by Urs Scherer. Negotiations started in 1968-70. Wild was to concentrate its efforts on the development and perfection of its stereo microscopes and ophthalmic instruments. Development costs were being reduced and channels of distribution were automatically improved, using established Leitz and Wild agencies world wide.

Dr. Max Kreis became the first Wild official in a management position at Leitz, together with the Leitz brothers. He exercised his powers for several years to supervise the company's restructuring to implement the various synergies anticipated as a result of the cooperation. In 1974, Wild purchased 51% of Leitz Wetzlar and was now totally in charge of its destiny. Max Kreis retired in 1978-79 and Urs Scherer became his successor in 1982. This year was significant because E. Leitz, Inc., New York was taken over from the Mann family by Wild after expiration of a contract which had lasted for 30 years.

Urs Scherer was not predisposed toward the optical or photographic industry, as his background was in cement. Nonetheless, he enjoyed his new environment until he too was replaced in 1987. A younger Schmidheiny generation was now taking charge. And, the company had been renamed Wild/Leitz in December 1988.

Since 1971 Leitz had been negotiating with Minolta to secure cooperation in the photographic market. Minolta was chosen because it did not manufacture microscopes. Cooperation was primarily sought in the camera manufacturing area. Patents and technical know-how were to be exchanged for mutual benefit as follows:

1. The new CL, "Peoples Camera", designed by Stein.

2. Delivery of components for the Leicaflex R3, Leicaflex R4 and R5.

A short time later (1990), Wild/Leitz merged with Cambridge Instruments and founded the so called "Leica" Group. It resulted in the dismantling of the Leitz U.S.A. sales organization, basically destroying the Leitz trademark for microscopes and scientific instruments in the U.S.A. in favor of the trademark for the "Leica" which, after all, referred only to a camera, the acronym for Leitz Camera.

As of 1985, the Leitz family did not have a successor for a management position. The family assets were almost exclusively tied up in the factory. The share owners (16 family members) had diverse views and interests and, as minority holders, were not able to exert much influence over the conduct of the business. Finally, an offer was made by the majority share holders to purchase the Leitz sales agencies in Germany, as well as the remaining shares held by Leitz family members.

The Leitz family dynasty, which had played a major part in the worldwide development of scientific instrumentation and photography during the last 150 years, was fading from view. Guenther Leitz died in 1969. Dr. Ernst Leitz III died in 1979. Dr. Ludwig Leitz died in 1992. Dr. Elsie Kuehn-Leitz died in 1985.

The Leica Camera is today manufactured and developed in its own facilities, Leica GmbH, Solms, about 10-12 miles from its old headquarters in Wetzlar. The decision to separate microscope manufacture from consumer products, such as Leica cameras, Leica lenses and accessories, field glasses, projectors and enlargers was made in 1987.

Postscript:

Brown & Sharp Manufacturing Company
Precision Park, 200 Frenchtown Road
North Kingstown, R.I. U.S.A.

acquired the

Leitz Metrology Division in 1990
Leitz Messtechnik GmbH
Siegmund Hiepe Strasse 2-12
D-35578 Wetzlar, Germany
Tel.: 49-6441-207-0